Exploring Plane and Solid Geometry

with THE GEOMETER'S SKETCHPAD® VERSION 5

Addressing the Common Core State Standards for Mathematics

Writers: Dan Bennett, Christopher Casey, Greg Clarke, Larry Copes, Deidre Grevious, Lynn Hughes, Rhea Irvine, Ross Isenegger, Nick Jackiw, Tobias Jaw, Amy Lamb, Paul Kunkel, Ann Lawrence, Andres Marti, Daniel Scher, Nathalie Sinclair, Scott Steketee, Kelly Stewart, Kevin Thompson

Reviewers: Karen Anders, Susan Beal, Janet Beissinger, Dudley Brooks, Greg Clarke, Gord Cooke, Larry Copes, Judy Dussiaume, Paul Gautreau, Shawn Godin, Paul Goldenberg, Lynn Hughes, Scott Immel, Ross Isenegger, Sarah Kasten, Cathy Kelso, Amy Lamb, Dan Lufkin, Aaron Madrigal, Linda Modica, Margo Nanny, Henry Picciotto, Nicolle Rosenblatt, Joan Scher, Dick Stanley, Tom Steinke, Glenda Stewart, John Threlkeld, Philip Wagreich, Ken Waller, Bill Zahner, Danny Zhu

Field Testers: Laura Adler, Joëlle Auberson, Vera Balarin, Kim Beames, Judy Bieze Caron Cesa, Heather Darby, Robin Glass, Cherish Hansen, Layne Hudes, Lynn Hughes, Scott Immel, Susan Friedman, Ann Lawrence, Vanessa Mamikuniam, Michelle Mancini, Michelle Moore, Margo Nanny, Kelly O'Keefe, Tina Pierorazio, Leslie Profeta, Dechelle Rasheed, Cheryl Schafer, Joan Scher, Kimberly Scheier, JoAnne Searle, Char Soucy, Jessie Starr, Ruth Steinberg, Nancy Stevenson, Terry Suetterlein, Mona Sussman, William Vaughn, Ethan Weker, Angie Whaley

Editors: Andres Marti, Elizabeth DeCarli

Contributing Editors: Rhea Irvine, Daniel Scher, Josephine Noah, Scott Steketee, Cindy Clements, Joan Lewis, Silvia Llamas-Flores, Kendra Lockman, Lenore Parens, Glenda Stewart, Kelly Stewart

Editorial Assistant: Tamar Chestnut

Production Director: Christine Osborne

Production Editors: Angela Chen, Andrew Jones, Christine Osborne

Other Contributers: Judy Anderson, Elizabeth Ball, Tamar Chestnut, Brady Golden, Ashley Kuhre, Nina Mamikunian, Marilyn Perry, Emily Reed, Ann Rothenbuhler, Juliana Tringali, Jeff Williams

Copyeditor: Jill Pellarin

Cover Designer: Suzanne Anderson

Cover Photo Credit: imagewerks

Printer: Lightning Source, Inc.

Executive Editor: Josephine Noah

Publisher: Steven Rasmussen

Key Curriculum Press
1150 65th Street
Emeryville, CA 94608
510-595-7000
editorial@keypress.com
www.keypress.com

ISBN: 978-1-60440-227-8
10 9 8 7 6 5 4 3 2 1 15 14 13 12 11

Contents

Downloading Sketchpad Documents

Getting Started

All Sketchpad documents (sketches) for *Exploring Plane and Solid Geometry in Grades 6–8 with The Geometer's Sketchpad* are available online for download.

- Go to www.keypress.com/gsp5modules.
- Log in using your Key Online account, or create a new account and log in.
- Enter this access code: EP68z537
- A Download Files button will appear. Click to download a compressed (.zip) folder of all sketches for this book.

The downloadable folder contains all of the sketches you need for this book, organized by chapter and activity. The sketches require The Geometer's Sketchpad Version 5 software to open. Go to www.keypress.com/gsp/order to purchase Sketchpad, or download a trial version from www.keypress.com/gsp/download.

Types of Sketches

Student Sketches: In many activities, students use a prepared sketch that provides a model, a simulation, or a complicated construction to investigate relationships. You will often use the student sketch to introduce an activity, guided by the Activity Notes. The name of a Student Sketch usually matches the activity title, and is referenced on the Student Worksheet for the activity, such as **Cube Nets.gsp.**

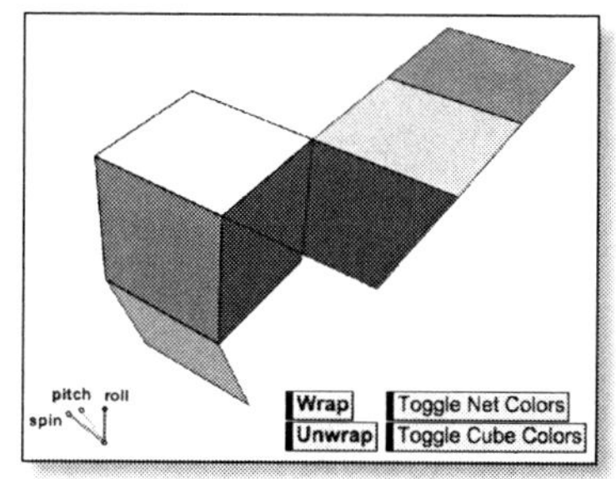

Presentation Sketches: Some activities include sketches designed for use with a projector or interactive whiteboard, either for a teacher presentation or a whole-class activity. Presentation Sketches often have action buttons to enhance your presentation. For activities in which students create their own constructions, the Presentation Sketch can be used to speed up, summarize, or review the mathematical ideas from the activity. The name of a Presentation Sketch always ends with the word "Present," such as **Polygon Pretenders Present.gsp.**

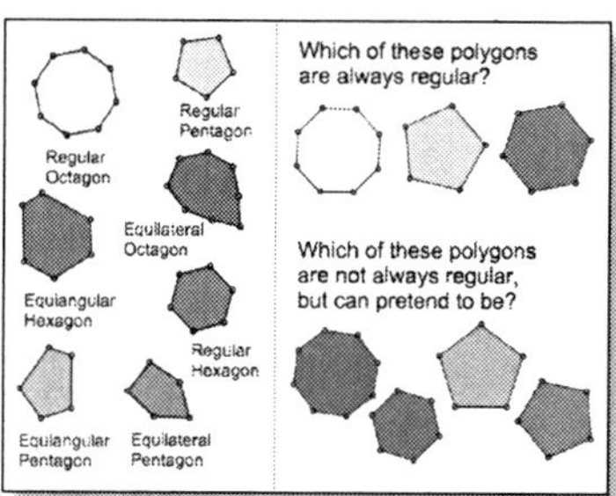

Sketchpad Resources

Sketchpad Learning Center

The Learning Center provides a variety of resources to help you learn how to use Sketchpad, including overview and classroom videos, tutorials, Sketchpad Tips, sample activities, and links to online resources. You can access the Learning Center through the Sketchpad's start-up screen or through the Help menu.

Help
- Learning Center
  - Welcome Videos
  - Using Sketchpad
  - Teaching with Sketchpad
- Reference Center
  - Objects
  - Tools
  - Menus
- Online Resource Center
  - Sample Sketches & Tools
  - Check for Updates
  - Technical Support
- Picture Gallery
- License Information...
- About Sketchpad...

The Learning Center has three main sections:

Welcome Videos

These videos introduce Sketchpad from the point of view of students and teachers, and give an overview of the big ideas and new features of Sketchpad 5.

Using Sketchpad

This section includes 12 self-guided tutorials with embedded videos, 70 Sketchpad Tips, and links to local and online resources.

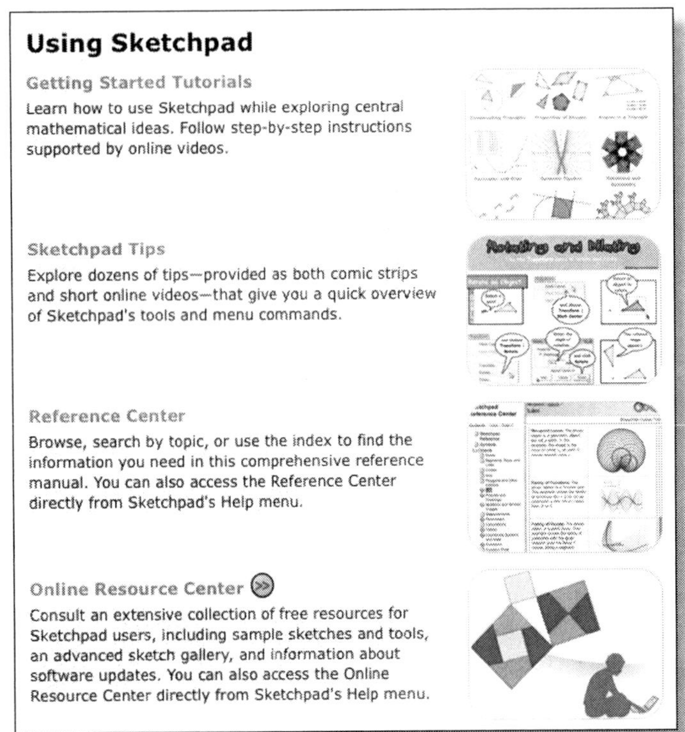

Teaching with Sketchpad

This section includes videos and articles describing how teachers make effective use of Sketchpad and how it affects their students' attitudes and mathematical understanding. There are over 40 sample activities, each with an overview, teaching notes, student worksheet, and sketches, that you can use with students to support your grade level, and curriculum.

Other Sketchpad Resources

Exploring Plane and Solid Geometry in Grades 6–8 with The Geometer's Sketchpad activities are an excellent introduction to Sketchpad for both students and teachers. If you want to learn more, Sketchpad contains resources for beginning and advanced users..

- **Reference Center:** This digital resource, which is accessed through the Help menu, is the complete reference manual for Sketchpad, with detailed information on every object, tool, and menu command. The Reference Center includes a number of How-To sections, an index, and full-text search capability.
- **Online Resource Center:** The Geometer's Sketchpad Resource Center (www.dynamicgeometry.com) contains many sample sketches and advanced toolkits, links to other Sketchpad sites, technical information (including updates and frequently asked questions), and detailed documentation for JavaSketchpad, which allows you to embed dynamic constructions in a web page.
- **Sketch Exchange:** The Sketchpad Sketch Exchange™ (sketchexchange.keypress.com) is a community site where teachers share sketches and other resources with Sketchpad users. Browse by keyword or topic for sketches that interest you, or ask questions and share ideas in the forum.
- **Sample Sketches & Tools:** You can access many sketches, including some with custom tools, through Sketchpad's Help menu. You can use some sample sketches as demonstrations, others to get tips and information about particular constructions, and others to access custom tools that you can use to perform special constructions. These sketches are also available under General Resources at the Sketchpad Resource Center (www.dynamicgeometry.com).
- **Sketchpad LessonLink™:** This online subscription service includes a library of more than 500 prepared activities (including those used in this book) aligned to leading math textbooks and state standards for grades 3–12. For more information, additional sample activities, or a trial subscription, go to www.keypress.com/sll.
- **Online Courses:** Key Curriculum Press offers moderated online courses that last six weeks, allowing you to immerse yourself in learning how to use Sketchpad in your teaching. For more information, see Sketchpad's Learning Center, or go to www.keypress.com/onlinecourses.
- **Other Professional Development:** Key Curriculum Press offers free webinars on a regular basis. You can also arrange for one-day or three-day face-to-face workshops for your district or school. For more information, go to www.keypress.com/pd.

Addressing the Common Core State Standards

The Common Core State Standards for Mathematics emphasize the importance of students gaining expertise with a variety of mathematical tools, including dynamic geometry® software such as The Geometer's Sketchpad. The Standards for Mathematical Practice specify that students should use technological tools to explore and better understand mathematical concepts.

> **5. Use appropriate tools strategically.**
>
> Mathematically proficient students consider the available tools when solving a mathematical problem. These tools might include pencil and paper, concrete models, a ruler, a protractor, a calculator, a spreadsheet, a computer algebra system, a statistical package, or dynamic geometry software. Proficient students are sufficiently familiar with tools appropriate for their grade or course to make sound decisions about when each of these tools might be helpful, recognizing both the insight to be gained and their limitations. They are able to use technological tools to explore and deepen their understanding of concepts. (Common Core State Standards for Mathematics, 2010, www.corestandards.org)

This collection of Sketchpad activities supports you in using dynamic geometry® software to address the Common Core State Standards for Mathematics for Grades 6 to 8. In addition to supporting the Standards for Mathematical Practice, these activities give students the opportunity to explore content from the domain of Geometry. As a set, they address the standard cluster statements below.

Grade 6: Geometry

- Solve real-world and mathematical problems involving area, surface area, and volume.

Grade 7: Geometry

- Draw, construct and describe geometrical figures and describe the relationships between them.
- Solve real-life and mathematical problems involving angle measure, area, surface area, and volume.

Grade 8: Geometry

- Understand congruence and similarity using physical models, transparencies, or geometry software.
- Understand and apply the Pythagorean Theorem.
- Solve real-world and mathematical problems involving volume of cylinders, cones and spheres.

Overview of Content

The content of each chapter is described below, as well as how the activities in that chapter correlate to the Common Core State Standards. All of these activities strongly correlate to the Standards for Mathematical Practice. The Geometer's Sketchpad is a mathematical tool that students can use strategically, allows them to model with mathematics, and motivates them to make sense of problems and persevere in solving them. In addition, the worksheets and activity notes are designed to help students reason abstractly and quantitatively, look for and make use of structure, and look for and express regularity in repeated reasoning.

Chapter 1: Angle Relationships

Students construct simple geometric models to explore types of angles and angle relationships formed by intersecting lines and by parallel lines intersected by a transversal. They also solve both mathematical and real-world problems involving angle measures.

Chapter 2: Triangle Properties

With triangles as the focus, students construct and describe geometric figures and describe the relationships among them. In addition to general triangle properties, including triangle congruence, students construct and investigate isosceles and right triangles and develop an understanding of the Pythagorean Theorem and its applications.

Chapter 3: Polygon Properties

Continuing with special quadrilaterals, and moving to general polygons, students construct and describe geometric figures, describe the relationships among them, classify shapes based on their properties, and develop their understanding of congruence.

Chapter 4: Area Formulas

Students develop methods for calculating the area of parallelograms, triangles, trapezoids, circles, and regular polygons. They understand that height is a perpendicular distance to the base, and solve real-world and mathematical problems involving area.

Chapter 5: Transformations

Students use Sketchpad's basic transformation commands to build their own models and develop their understanding of congruence and similarity. The activities start with translations, rotations, and reflections, focusing on the coordinate grid, and end with dilations, which preserve similarity but not congruence.

Chapter 6: Three-Dimensional Solids and Volume

These activities start with a focus on spatial visualization skills, building to activities in which students deconstruct a solid into its two-dimensional net and use this to find the surface area. The prism and pyramid models can be taken to their limits to investigate cylinders and cones. Other models develop the concept of volume and lead toward solving real-world and mathematical problems about surface area and volume.

1

Angle Relationships

Measure by Degrees: Types of Angles

Students create, drag, and measure angles. In the process, they are introduced to Sketchpad and review terms associated with angles. Students use Sketchpad tools and create a simple animation.

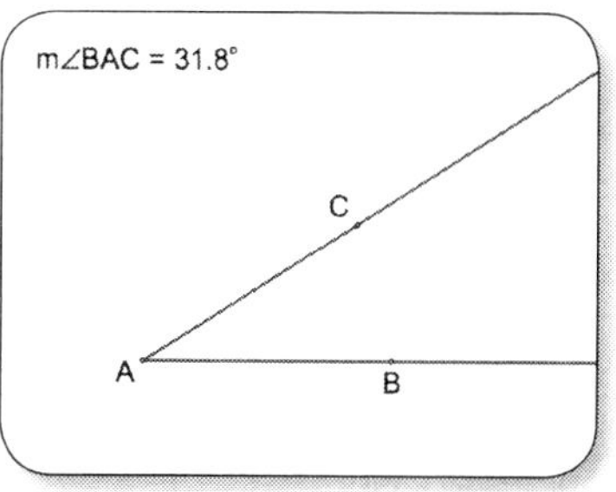

Racquetball: Using a Protractor

Students explore a model that represents a ball bouncing off a wall. They predict where to place their racquet to hit the ball after the ball bounces. Then students use a Sketchpad protractor to measure incoming and outgoing angles.

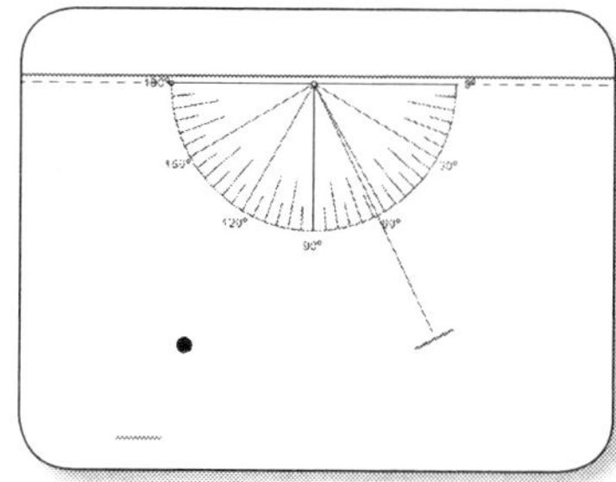

When Lines Meet: Conjectures about Angle Pairs

Students construct geometric models and manipulate them to investigate adjacent, vertical, supplementary, and complementary angles. Students make and test conjectures about angle relationships, find missing angle measurements, and use geometric terminology for lines and angles.

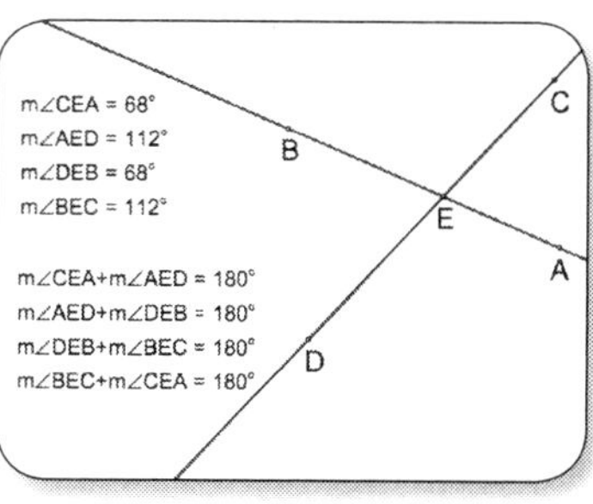

Double Cross: Angles Formed by a Transversal

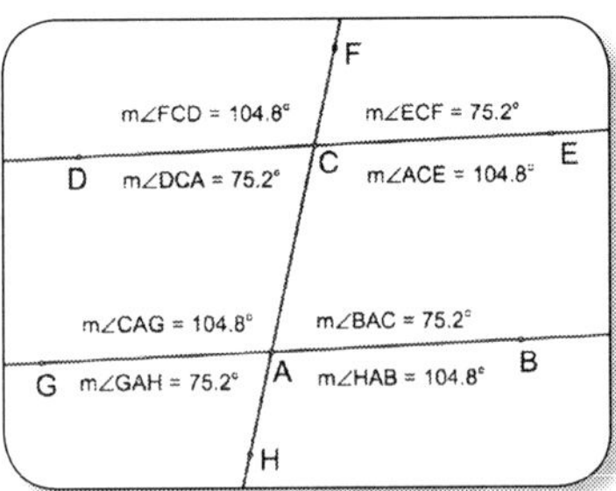

Students explore the angles formed when a transversal intersects two parallel lines. They find pairs of congruent and pairs of supplementary angles. In the process, they review terms for the angles formed and become more familiar with Sketchpad.

Amazing Angles: Finding Transversal Angle Pairs

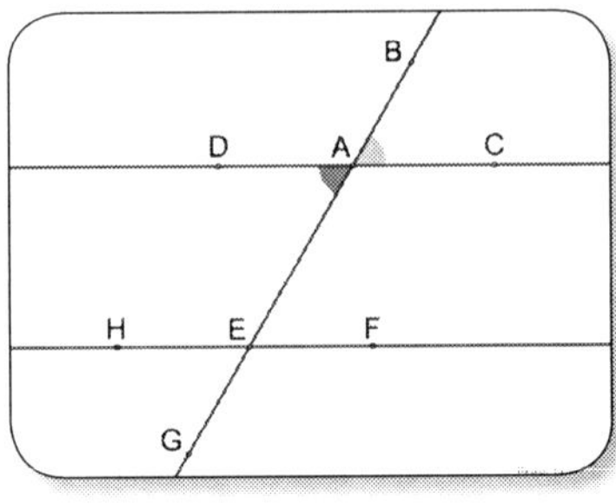

Students investigate the relationships among the angles created by a transversal intersecting two parallel lines. Students focus on the vocabulary used to describe the congruent and supplementary angles and solve problems that require them to find paths from one of the eight angles to another.

INTRODUCE

Project the sketch for viewing by the class. Expect to spend about 10 minutes.

1. Explain, ***Today you're going to use Sketchpad to explore angles and angle measures. An angle is sometimes defined as two rays that share an endpoint. Two segments with a common endpoint can also determine an angle. An angle can be named after three points or one. The angle measure is different from the angle.***

2. ***I'll demonstrate how to construct and measure an angle, and then you'll construct and measure angles on your own.*** Start with a new sketch and enlarge the document window so it fills most of the screen. As you demonstrate, make lines thick and labels large for visibility. First, model the angle construction in worksheet steps 1–5. Then model how to find an angle measure in worksheet step 8. Here are some tips.

   - In worksheet step 1, explain, ***Angles are measured in degrees. For today's exploration we will want to measure in degrees to the nearest tenth.*** Model how to choose **Edit | Preferences | Units** and set Angle Units to **degrees** and Angle Precision to **tenths**.

To change a label, choose the **Text** tool and double-click the label. In the dialog box, type a new label and click **OK.**

   - In worksheet step 2, construct $\overrightarrow{AB}$. Model choosing the **Ray** tool. Labels for the points will automatically appear if you start with a new sketch and choose **Edit | Preferences | Text** and check **Show labels automatically: For all new points.** Model how to add or change labels using the **Text** tool.

   - Once the points are labeled correctly, click on the two points in this order: endpoint of the ray (point *A*) and the other point on the ray (point *B*). Explain that students will wait to label the points until the construction is complete; you are labeling now so that you can refer to the points as you demonstrate.

   - In worksheet step 3, use the **Ray** tool to construct $\overrightarrow{AC}$. ***When endpoint A is selected, the point will be highlighted. When it is highlighted, click point A and drag to construct the new ray, the second side of the angle.*** Make sure the third point is labeled *C*.

   - In worksheet step 4, model dragging the three points around with the **Arrow** tool. Emphasize the importance of this drag test to be sure that the two rays share a common endpoint.

   - Make sure students know how to use **Edit | Undo** if they make a mistake.

- Now model how to measure the size of the angle. ***The points of the angle must be selected in the correct order with the vertex as the second point.*** Select the three points, with point *A* second, and then choose **Measure | Angle**. Read the measure aloud, noting that the degrees are in tenths as desired. ***The* m *in front of the angle stands for "the measure of."***
- In worksheet step 8, after measuring the angle, drag point *C* to demonstrate that the angle measure updates automatically as the size of the angle changes.

3. If you want students to save their work, demonstrate choosing **File | Save As,** and let them know how to name and where to save their sketches.

DEVELOP

Expect students at computers to spend about 20 minutes.

4. Assign students to computers. Distribute the worksheet. Tell students to work through step 21 and do the Explore More if they have time. Encourage students to ask their neighbors for help if they are having difficulty with the construction.
5. Let pairs work at their own pace. As you circulate, here are some things to notice.
   - For worksheet steps 6 and 7, look to be sure students understand what the vertex of the angle is. Restate, if needed, that the vertex is the point where the two sides of the angle meet. Introduce or review naming an angle using the angle symbol and the label of the vertex.
   - In worksheet step 10, students should notice that the angle measure increases and then begins to decrease once it passes 180°. Sketchpad acts like most protractors, so it does not measure reflex angles.
   - For step 12, students may not describe the rays as perpendicular; instead they may say the angle looks like "a corner." Encourage students to use geometric terms in their descriptions.
   - Since worksheet step 16 was not modeled, some students may not successfully construct the circle centered at point *A*. Tell students to be sure point *A* is highlighted before they click and drag to construct the circle. Mention the drag test to check that point *A* remains the center of the circle.

- For step 17, check that students select only the circle and point *C*.
- For step 21, tell students that they can look back to steps 2–5 for help, but they should choose the **Segment** tool, not the **Ray** tool, to construct the angle this time. Notice students who are unsure whether the size of the angle depends on the lengths of the angle's sides. Are they able to convince themselves that this is not so by lengthening the sides while decreasing the angle measure, or by shortening the sides while increasing the angle measure?
- The Explore More helps students develop visualization and estimation skills. They make angles of different sizes, classify each angle, and estimate the angle measure before measuring to get immediate feedback.

SUMMARIZE

Project the sketch. Expect to spend about 15 minutes.

6. Gather the class. Students should have their worksheets. Open **Measure by Degrees Present.gsp** and use page "Angle Measure" as needed. Begin the discussion by checking that students understand how an angle is formed. ***What did you use to construct the sides of your angle in steps 2–5? In step 21? Based on your constructions, how would you define* angle?** Write, "An angle is" on chart paper or other display and help the class create a clear definition. Here is a sample definition: An angle is a figure formed by two rays, or segments, that share a common endpoint called the *vertex*. Review the terms *ray, segments, endpoint,* and *vertex* as needed.
7. Review worksheet steps 6 and 7. ***What are the different ways to name an angle?*** Write "An angle can be named" on the display. Students should reply that an angle can be named by its vertex, such as $\angle A$ for the first construction, or by three points: one point on one side of the angle, the point at the vertex, and one point on the other side, such as $\angle BAC$ or $\angle CAB$. Be sure students understand that the vertex is always the second point when naming an angle using three points. Introduce the angle symbol if students are not familiar with it.
8. Discuss worksheet steps 11–15 with the class. Have students define the different types of angles: right, straight, acute, and obtuse. Add the definitions to the chart paper. (You don't need to introduce *reflex angle* at this time.) Then ask students to classify the different angles they see in their environment. Students may suggest the following ideas.

Right angles are found in the corners of windows, doors, walls, desks, and my worksheet.

Where the chair leg meets the seat is a right angle.

The hands of the clock make an acute angle when the time is 10:00. The hands make an obtuse angle when the time is 10:15.

The roof line of the school forms an obtuse angle.

The bottom of the white board is a straight angle.

9. Discuss worksheet step 20. Have students share how they arrived at their answers. ***If a complete revolution forms a circle, how many degrees are in a circle?*** Talk about how many degrees are in one-fourth and one-half revolutions. ***What angles do these represent?*** [Right and straight]
10. Have students discuss their responses to worksheet step 21; invite a volunteer to the computer to demonstrate. Be sure students understand that the length of an angle's sides is not related to the measure of its angle, a key misconception that many students have.
11. If students had time for the Explore More, ask them to share their findings.

ANSWERS

6. The vertex of the angle is point A. The name of the angle using just the vertex is $\angle A$.
7. $\angle BAC$, $\angle CAB$
10. The smallest angle measure is 0°. The greatest is 180°.
11. An angle with measure 0° is a single ray.
12. An angle with measure 90° has perpendicular sides.
13. An angle with measure 180° is a straight line.
14. Drawing shows acute angle.
15. Drawing shows obtuse angle.
20. The greatest angle measure would be 360°.
21. If an angle has segments for sides, it is possible to lengthen the sides while actually decreasing the angle measure.
23. Angle sketches and measures should match.

Measure by Degrees

Name:

In this activity you'll explore angles and angle measures.

CONSTRUCT

1. In a new sketch, choose **Edit | Preferences** and go to the Units panel. Set Angle Units to **degrees** and Angle Precision to **tenths.** Click **OK.**

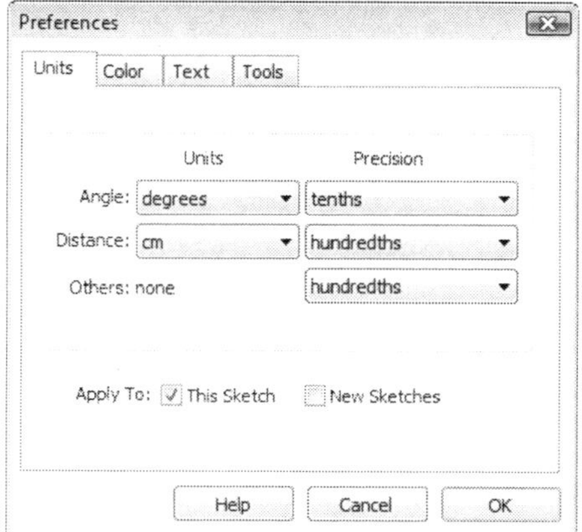

2. Construct $\overrightarrow{AB}$.

3. Now construct $\overrightarrow{AC}$ with the same endpoint *A*.

4. Drag each of the three points to make sure the two rays share a common endpoint (point *A*).

   If you need to, choose **Edit | Undo** to back up one or more steps.

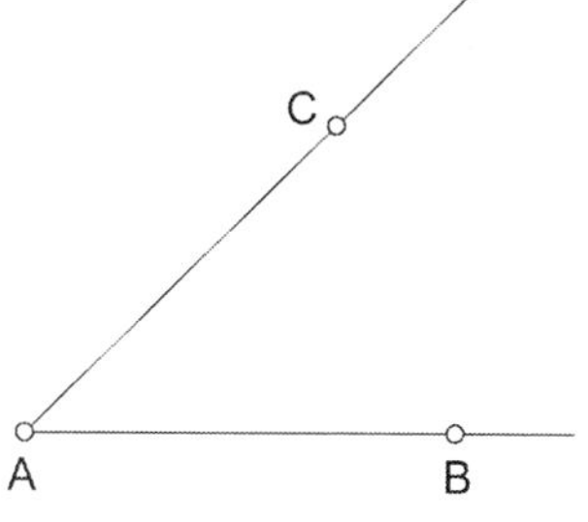

5. If necessary, use the **Text** tool to display the point labels. Change them to match the figure.

   To change a label, double-click the label. In the dialog box, type a new label and click **OK.**

EXPLORE

6. Two rays with a common endpoint form an angle. The common endpoint is called the *vertex* of the angle. Sometimes angles are named just by their vertex. What is the vertex of the angle you just made? ____________

   What is the name of your angle using just the vertex? ∠ ____________

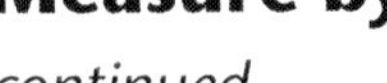

Measure by Degrees

continued

7. Angles can also be named after three points: a point on one side, the vertex, and a point on the other side. The vertex is always named second. What are two other possible names for the angle you just made?
   ∠ ____________ and ∠ ____________

8. Select, in order, points *B*, *A*, and *C*. Choose **Measure | Angle.** When you select an angle in Sketchpad, the vertex must be your second selection.

9. Drag point *B* or point *C* and observe how the angle measure changes.

10. What is the smallest possible angle measure? ____________
    What is the greatest possible angle measure? ____________

11. Drag a point on your angle until the angle's measure is as close to 0° as possible. Describe this angle.

12. Drag a point on your angle until the angle's measure is as close to 90° as possible. Describe this angle.

13. Drag a point on your angle until the angle's measure is as close to 180° as possible. Describe this angle.

14. An acute angle has a measure between 0° and 90°. Drag a point on your angle to make it acute. Sketch an example of an acute angle.

15. An obtuse angle has a measure between 90° and 180°. Drag a point on your angle to make it obtuse. Sketch an example of an obtuse angle.

16. Construct a circle centered at point *A* but not attached to any other points in your sketch.

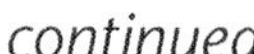

17. Select the circle and point *C*.

    Then choose **Edit | Merge Point to Circle.**
    Point *C* will attach itself to the circle.

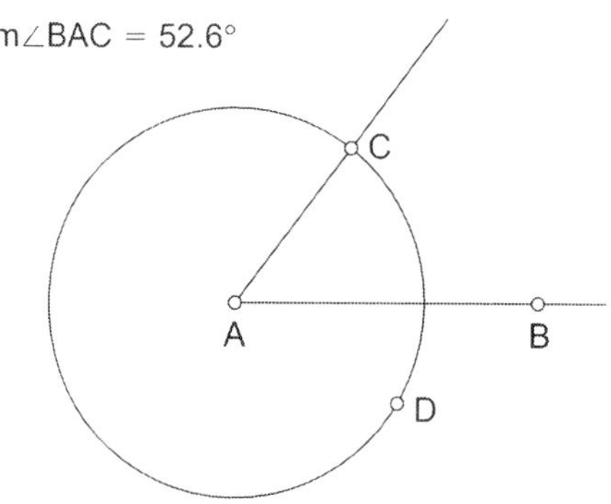

18. Now you will make an action button to animate point *C* around the circle.

    Select point *C* and choose
    Edit | Action Buttons | Animation.

    Set Speed to **slow** and click **OK.**

19. Press *Animate Point.* Watch the angle measurement. Stop the animation by pressing the button again.

20. Because of the way Sketchpad measures angles, when one angle side rotates by more than one-half a revolution from the other angle side, the angle measure starts to decrease. Suppose the angle measure kept increasing until the angle side had completed one whole revolution.

    What would be the greatest angle measure? ____________

21. Construct a new angle using the **Segment** tool.

    Does the measure of the angle depend on the lengths of its sides? Explore this question in Sketchpad. Then give your answer and explain your reasoning.

EXPLORE MORE

22. Get ready to try estimating the sizes of angles.

    Using the angle you made in step 21, measure the angle.

    With the measure selected, choose **Edit | Action Buttons | Hide/Show.**
    Now you can press this button when you want to hide or show the angle measurement.

23. Drag the endpoints of the segments to make new angles. Estimate the angle sizes and then measure them. Record your work in the table. Remember to hide the angle measurement before you estimate.

Type of Angle	Estimate	Measure

INTRODUCE

Project the sketch for viewing by the class. Expect to spend about 5 minutes.

1. If most of your students know how to measure angles with a protractor, you may not need to introduce this activity. Skip to step 3.
2. To introduce or review using a protractor, open **Racquetball.gsp** and go to page "Practice." Have a volunteer manipulate the pointer according to your instructions.
   - Move the protractor until the center of its straight side is at the angle's vertex. (You move the protractor by dragging an *unmarked* point on its edge.)
   - Rotate the protractor by dragging the red point at 180°. Put the 0 point on one side of an angle. Make sure that the other side of the angle also goes through the protractor's edge. You can change the direction of the protractor from measuring from 0° to 180° to measuring from 180° to 0° by pressing *Change Direction.*
   - Read off what number the other side goes through. (If the side goes between marked points, help students decide how to estimate the value.)

DEVELOP

Expect students at computers to spend about 20 minutes.

3. Explain, ***Your goal is to find patterns in the measurements of angles formed when a ball bounces against a wall.*** Assign students to computers and tell them where to locate locate **Racquetball.gsp.** Distribute the worksheet. Tell students to work through step 9 and do the Explore More if they have time. Suggest that students help each other if they're having difficulties with Sketchpad.
4. Let students work at their own pace. Here are some things to notice.
   - On page "One Wall," students are experiencing the fact that the incoming and outgoing angles are congruent. Raise the question, ***How can you predict where the ball will go?***
   - In worksheet step 3, you may need to help individual students review how to calculate a percent.
   - Students are not asked to show and measure angles until page "Analyze."

- For students struggling to get the protractor in the correct position to measure an angle ask, ***What does pressing* Change Direction *do?***
- In worksheet step 7, students may see various patterns. As needed, help them see that the incoming and outgoing angles have the same measure.
- In worksheet step 8, some students who can measure given angles may have difficulty drawing an angle of a given measure. Ask, ***Do you know one side of the angle?*** [The dashed line] ***Do you know the vertex? Can you position the protractor on the known side and the vertex? Where will the other side go?*** Help students use the **Ray** tool to draw the path of the bounced ball from the vertex through the desired value on the protractor. ***Where will you put the red racquet to intercept the ball?***
- Note a student, or perhaps more than one, who has an interesting description in worksheet step 9 to present to the class.

SUMMARIZE

Expect to spend about 5 minutes.

5. Gather the class. Students should have their worksheets with them. Have one or more students present ideas about the process in step 9. Reach a class consensus that the incoming and outgoing angles between the path and the wall have the same measure. You may want to introduce the word *congruent* for such angles. Mathematicians don't say that the angles themselves are equal; rather, the angles are congruent if their measurements are equal numbers. Whether or not your class is ready to make this distinction, you should use the terminology correctly.

   If a student should happen to refer to the *angle of incidence* and the *angle of reflection*, you might mention that these terms usually refer to the angles with the perpendicular to the wall, not to the wall itself. They are the complements of the angles from the wall.

6. ***What other questions can you ask about angles and reflection? You can ask a question even if you don't know the answer.*** Questions of mathematical interest include these.
   - Is there an easier way to measure angles than with a protractor?
   - Why doesn't the measurement change when you change the size of the protractor?

- Why do the incoming and outgoing angles have the same measure?
- What if the wall is curved?
- Why are angles measured in degrees? Are there other units of angle measurement?

EXTEND

Students who are having difficulty measuring angles might benefit from working alone or in pairs on page "Practice."

ANSWERS

3. Answers will vary.
4. Answers will vary.
6. Measurements will vary but should add to 180°. Two of the three angles should have the same measurement.
7. The incoming and outgoing angle measures are equal. The sum of the angle measures is 180°.
9. Answers will vary. Sample Answer: I made an outgoing angle of the same measure as the incoming angle.
10. At each wall, the incoming and outgoing angles have the same measure. The first and last paths of the ball are parallel.

Racquetball

Name:

In this activity you'll explore the angles formed by the path of a racquetball as it bounces off a wall.

EXPLORE

1. Open **Racquetball.gsp** and go to page "One Wall." You'll see a small ball, a blue segment representing your opponent's racquet, and a red segment representing your racquet.

2. When you press *Serve,* your opponent is going to serve the ball against the wall. Using the **Arrow** tool, you will move the red racquet to hit the ball after it bounces off the wall. You can increase the challenge by making the racket smaller.

3. Press *Serve* and try to hit the ball. Then press *New Ball* and try again. Repeat at least a few times.

   How many times did the opponent serve the ball? __________

   How many times did you hit the ball? __________

   What percentage of the balls did you hit? __________

4. Go to page "Anticipate." This time the ball will go only to the wall and stop. Move the red racquet to a location that you think will let you hit the ball and then press *Bounce.* Repeat at least a few times.

   How many times did the opponent serve the ball? __________

   How many times did you hit the ball? __________

   What percentage of the balls did you hit? __________

5. How can you determine the path of the ball?

6. Go to page "Analyze." Press *Show Path* and press *Show Protractor.* Measure the three angles formed by the path, and record your measurements. Then press *New Ball* and repeat.

7. What patterns do you see in the angle measurements?

8. Go to page "Predict." Press *Show Path Before Bounce* and press *Show Protractor.* Use the protractor and the **Ray** tool to draw the path the ball will travel after bouncing. Put the red racquet on the path and then press *Serve.*

9. If you did not hit the ball, go back to step 6. Otherwise, explain how you determined the path the ball would travel after bouncing.

EXPLORE MORE

10. Go to page "Two Walls." Now you see a corner of the court. The ball will bounce off two walls. Try to hit it with the red racquet. Can you anticipate where the ball will go? Show the path of the ball. What angles have the same measure? What other patterns do you notice?

When Lines Meet: Conjectures About Angle Pairs

ACTIVITY NOTES

INTRODUCE

Project the sketch for viewing by the class. Expect to spend about 10 minutes.

1. Inform students, ***Today you're going to use Sketchpad to look for relationships among lines and angles, something early mathematicians did long ago.***
2. Open Sketchpad and enlarge the document so that it fills most of the screen.

   Model the construction in worksheet steps 1–4. Point out that four angles are formed and that the point of intersection is the vertex for all four angles.

   Model the dragging students will do in worksheet step 7. Alert students to keep point *E* between points *A* and *B* and between points *C* and *D*. If students have not used **Edit | Undo** when they make a mistake, demonstrate it now.

   Introduce the terminology for the angles created: *vertical angles* (angles that are opposite each other) and *adjacent angles* (angles that share a common side without overlapping).

   Students will also construct a ray, measure angles, and use the Calculator. If these skills are new, model them as well.
3. ***You will construct lines, look for relationships between angles, make some conjectures, and test them.*** Review the geometric notation for lines, rays, and angles that students will see on the worksheet. Explain that knowing and using the terms and notation will help the class discuss their findings.
4. Show students how to add a new blank page using **File | Document Options.**

DEVELOP

Expect students at computers to spend about 25 minutes.

5. Assign students to computers. Distribute the worksheet. Tell students to work through step 25 and do the Explore More if they have time. Encourage students to ask a neighbor for help if they have questions about using Sketchpad.

Exploring Vertical and Adjacent Angles

6. Let students work at their own pace. As you circulate, notice how they investigate vertical and adjacent angles, using their models

of intersecting lines. Here are some things to ask yourself as you observe.

- Is the student taking a visual approach, noticing angles that appear to be equal in size?
- Notice students taking an approach based upon mathematical attributes of the lines, such as, *The adjacent angles fit on a straight line, which is also a straight angle. We know that its measure is 180°.*
- Are students dragging a point to see whether a relationship they propose holds true for other positions of the lines?
- If students have observed a relationship for a pair of angles, have they looked to see whether there are other angle pairs that have the same relationship?

7. If any students are having difficulty looking for relationships, offer prompts that help them to investigate on their own. (Make a note about these students and visit them again when they have measured the angles in worksheet step 8.)

   What do you notice about angles that are vertical—opposite each other—like ∠AED and ∠BEC?

   What do you notice about the size of the angles when you drag point B?

   What do you observe about angles that are adjacent—next to each other—like ∠AED and ∠DEB?

   If students identify a couple of angle pair relationships and stop investigating, let them know that early mathematicians found more angle pairs.

Measuring Angles to Test Conjectures

8. In worksheet steps 8 and 9, students measure all four angles and use the displayed measurements to test their conjectures. This is an opportunity for students who did not see some relationships previously to recognize them. Some pairs may need to be reminded that to measure an angle, they need to select the vertex of the angle second.

9. In worksheet steps 10 and 11, if some students are unsure about conjecturing, you may want to check in with the class so these students can hear others' conjectures.

If this is the first time students have used Sketchpad's Calculator, make sure they understand that to calculate, they will click on measurements and also use the keys on the Calculator's keypad.

10. In worksheet step 12, note how many calculations students carry out. Some may be content with the calculation for one pair of adjacent angles. Other students may feel that they need two or four calculations to be convinced that any two adjacent angle measures sum to 180°. Note students whom you may want to call on during the summarizing discussion.

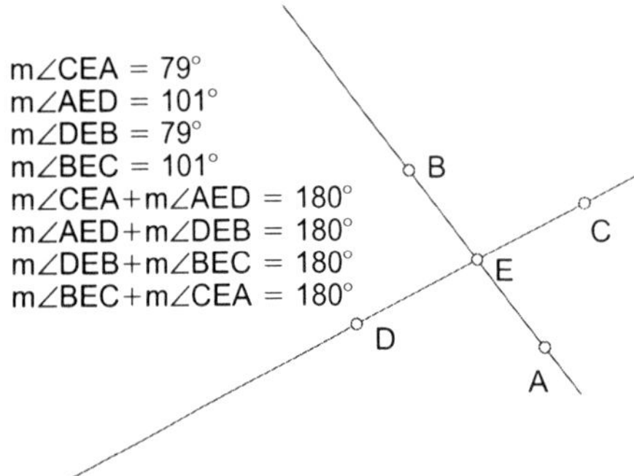

11. Worksheet step 13 provides an opportunity for students to become aware of the case in which all four angles formed by the intersecting lines are congruent. Take note of students who recognize that the lines are perpendicular. If some students don't see this, don't tell them; let this arise in the summarizing discussion.

m∠CEA = 90°
m∠AED = 90°
m∠DEB = 90°
m∠BEC = 90°
m∠CEA+m∠AED = 180°
m∠AED+m∠DEB = 180°
m∠DEB+m∠BEC = 180°
m∠BEC+m∠CEA = 180°

B E C D A

Exploring Supplementary and Complementary Angles

Don't use the terms *supplementary* or *complementary* yet. Let students use their own language as they work. You will make the connection to mathematical language in the summarizing discussion.

12. In worksheet steps 14–19, students use the same process they used to explore intersecting lines, this time to investigate supplementary angles.

13. In worksheet steps 20–25, students investigate complementary angles. This time they are asked to think how they can use Sketchpad to test their conjectures. If students aren't sure what they need to do, encourage them to think about their constructions and work with the previous models of lines and angles.

14. If time allows, have students do the Explore More, in which they are asked to determine the missing angle measurements given intersecting lines and one angle measurement. If students don't have time to do this on their own, they will have time as a class in the summarizing discussion.

SUMMARIZE

Project the sketch for viewing by the class. Expect to spend 10 minutes.

15. Bring the class together. Students will need their worksheets. Project **When Lines Meet Present.gsp.** Invite students to share their discoveries, using the projected sketch to illustrate their ideas. ***What did you find most interesting?***

16. Let students lead the discussion, deciding what to share. During the discussion, introduce new geometric terminology for the angle pairs as appropriate: *complementary* and *supplementary.*

    As students discuss page "Intersecting Lines," their discussion should bring out these ideas.

    - When two lines intersect, two pairs of vertical angles are formed.
    - The measures of vertical angles are equal. These angles are congruent.
    - Four pairs of adjacent angles are formed.
    - The sum of the measures of each pair of adjacent angles is 180°.

    As students discuss page "Straight Line," their discussion should bring out these ideas.

    - Angle *ACB* is a straight angle. Its measure is 180°.
    - The sum of the measures of the adjacent angles is 180°.
    - Angles *ACD* and *DCB* have the same relationship as the adjacent angles on page "Intersecting Lines"—their non-adjacent sides lie on a line, and the sum of their measures is 180°.

    As students discuss page "Perpendicular Lines," their discussion should bring out the idea that the sum of the measures of the adjacent angles will always be 90°.

17. Include discussion of the Explore More task. If some students had time for this on their own, have them explain how they found the missing angle measurements, using page "Explore More 1" of the sketch. Do one or more examples, dragging a point to change the displayed measurement, letting students find the missing measurements, and checking using the button.

    Next, go to page "Explore More 2." ***What is the missing angle measurement?*** Have students explain how they know the missing measurement. Do one or two more examples.

    Go to page "Explore More 3" and follow the same process.

18. You may want to provide an illustration of two intersecting lines and have students write individually in response to this prompt.

    When two lines intersect, how many angles do you need to measure in order to know the measure of all four angles? How would you explain this to someone who didn't know?

ANSWERS

Students can describe each angle in two different ways. For instance, $\angle BEC$ is the same as $\angle CEB$. For clarity, these answers show only one of the two ways, but either way should be considered correct.

5. There are two pairs of vertical angles: $\angle BEC$ and $\angle AED$; $\angle CEA$ and $\angle DEB$.
6. There are four pairs of adjacent angles: $\angle BEC$ and $\angle CEA$; $\angle CEA$ and $\angle AED$; $\angle AED$ and $\angle DEB$; $\angle DEB$ and $\angle BEC$.
7. Answers will vary. Students may see that the vertical angles ("opposite each other") appear to have equal measures ("are the same size"), and also that there are pairs of adjacent angles ("next to each other") whose measures sum to a straight angle ("add to 180°").
9. Answers will vary.
10. Students should propose that the measures of vertical angles are equal.
11. Students' answers should include that the sum of the measures of two adjacent angles is 180°.

12. Calculating the sum of the measures of any two adjacent angles will verify that the result is 180°. Students should use this result to reflect on their conjecture in step 11.

13. Wording will vary. For adjacent angles to be congruent, they must each measure 90°. If one of the four angles measures 90°, then all four angles must measure 90°. The lines are perpendicular.

18. The sum of the two angles is equal to a straight angle. Therefore, the sum of the measures of the two angles is 180°.

19. Measurements will vary. The calculation should result in a sum of 180°, which will verify or challenge the conjecture made in worksheet step 18.

24. The two angles together form a right angle. Therefore, the sum of the measures of the two angles is 90°.

25. Students should explain that they put a point *D* on the line perpendicular to line *AB*, and measured angles *BAC* and *CAD*. Using the Calculator, they found the sum of the measures of these adjacent angles. The calculation results in a sum of 90°, verifying or challenging the conjecture made in worksheet step 24.

26. Answers will vary. One missing angle measure (the vertical angle) will be equal to the displayed measurement. The other two missing angle measurements will be equal to each other; the measure of either is found by subtracting the displayed measure from 180°.

When Lines Meet

Name:

In this activity you will make some discoveries about angle relationships.

CONSTRUCT

1. In a new sketch, press and hold the **Segment** tool until a menu pops out. Choose the **Line** tool from the menu.

2. Construct two lines so that they intersect.

3. Construct a point at the intersection.

4. Click the **Text** tool on the points to show their labels. To make your labels look like those in this picture, click first on point *A*, then on *B*, and so forth.

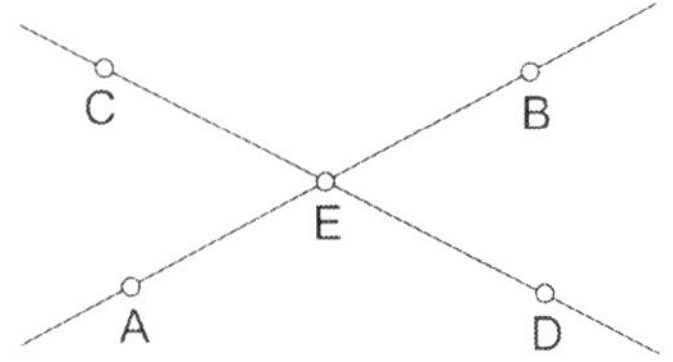

5. How many pairs of vertical (opposite) angles are there? Name them.

6. How many pairs of adjacent angles are there? Name them.

EXPLORE

7. Drag the points and look for relationships among the angles. Make sure to keep point *E* between points *A* and *B* and between points *C* and *D*.

   Describe any angle relationships you think you see.

8. Now you will measure $\angle CEA$, $\angle AED$, $\angle DEB$, and $\angle BEC$.

   To measure an angle, select three points, with the vertex second. Then choose **Measure | Angle.**

9. Drag each of the points and watch the measurements.

   Do the measurements confirm the angle relationships you saw?

   Do you notice anything new? Has anything changed your thinking?

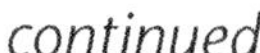

10. Write a conjecture about the measures of vertical angles.

11. Write a conjecture about the measures of two adjacent angles.

12. Now you will use the Calculator to test your conjecture in the last step. Choose **Number | Calculate** and use Sketchpad's Calculator to enter a calculation corresponding to your conjecture about adjacent angles. Click on a measurement to insert it into a calculation. Use + on the Calculator keypad to find sums.

    What did you find out?

13. Drag a point until the adjacent angles are congruent.

    What do you notice about the angles? About the lines?

14. Now you'll create a new page. Choose **File | Document Options.** Click Add Page and choose **Blank Page** from the drop-down menu.

15. Construct $\overleftrightarrow{AB}$.

16. Construct point *C* on the line. Label all three points.

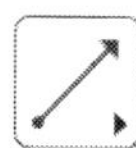

17. Press and hold the **Line** tool in the Toolbox, and choose the **Ray** tool from the menu that pops out. Construct $\overrightarrow{CD}$ and label point *D*.

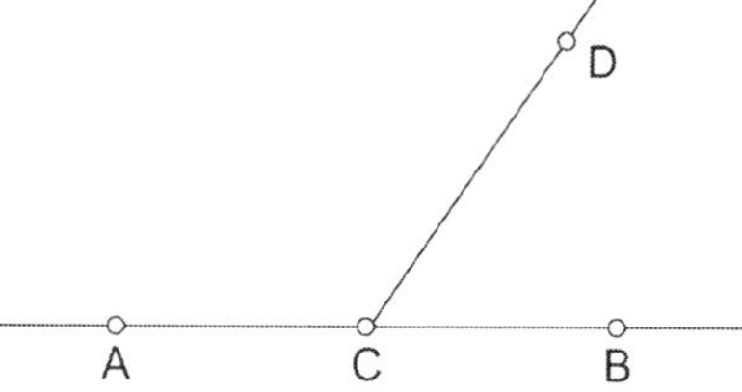

18. Drag point *D*. Write a conjecture about $\angle BCD$ and $\angle DCA$.

19. Measure angles and use the Calculator to test your conjecture.
    What did you find out?

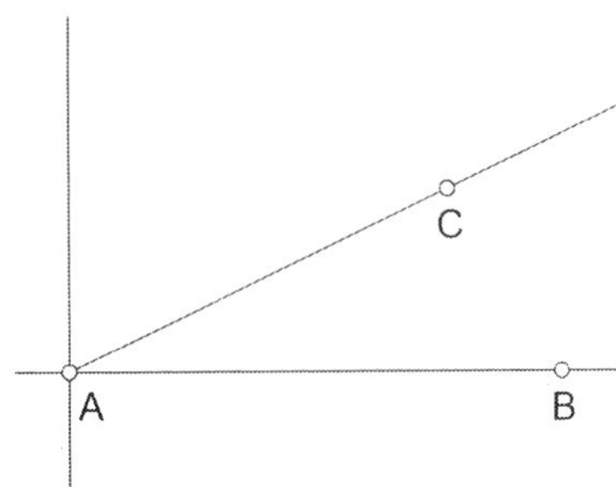

20. Create another blank new page.
21. Construct $\overleftrightarrow{AB}$ and label the points.
22. With point *A* and $\overleftrightarrow{AB}$ selected, choose **Construct | Perpendicular Line.**
23. Construct $\overrightarrow{AC}$ as shown.
24. Write a conjecture about the angles formed by the perpendicular lines and $\overrightarrow{AC}$.
25. How can you use Sketchpad to test your conjecture? What do you find out?

EXPLORE MORE

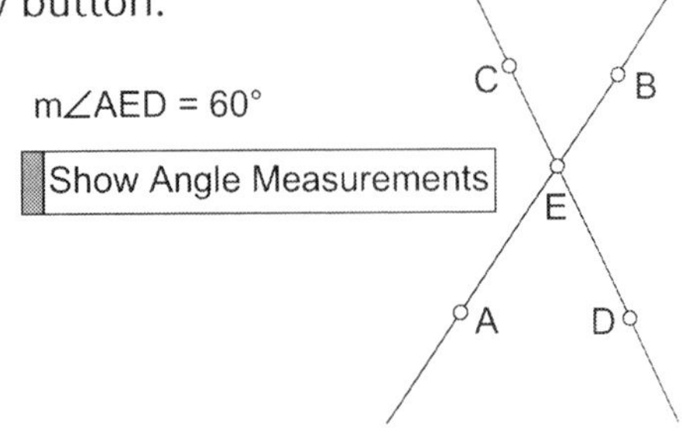

26. Go back to the first page. You will make a Hide/Show button.
    Select any three measurements.
    Choose **Edit | Action Buttons | Hide/Show.**
    Press the button that results. It hides the three measurements.
    Drag a point so that the angle measurement on screen is 60°.
    What are the measures of the other three angles?

    Press *Show Angle Measures* to check your thinking.
27. To test your understanding, repeat the process: Hide the measurements, change the angles, figure out the missing angle measures, and press the button to check.

Double Cross: Angles Formed by a Transversal

ACTIVITY NOTES

INTRODUCE

Project the sketch for viewing by the class. Expect to spend about 10 minutes.

1. Open Sketchpad and enlarge the document window so it fills most of the screen. Explain, ***Today you are going to use Sketchpad to explore the relationships among the angles formed when two parallel lines are intersected by a third line, called a* transversal. *I will demonstrate how to construct two parallel lines and a transversal and how to measure one angle formed by their intersection. Then you'll complete the same construction, but you'll measure all the angles formed and make conjectures based on your observations.***

2. As you demonstrate, make lines thick and labels large for visibility. First, model the construction in worksheet steps 1–8. Then model how to find an angle measure in worksheet step 9. Here are some tips.

Labels for points will automatically appear if you start with a new sketch and set **Edit | Preferences | Text** to **Show labels automatically: For all new points.**

To change a label, double-click the label. In the dialog box, type a new label and click **OK.**

- Follow worksheet steps 1–4 to construct $\overleftrightarrow{AB}$ and a line parallel to $\overleftrightarrow{AB}$ through point *C*. Drag all three points to show that the lines always remain parallel.
- In worksheet step 5, construct $\overleftrightarrow{CA}$. Drag points *C* and *A* to verify that the lines intersect at those points. Tell students to use **Edit | Undo** if they make a mistake in their constructions.
- Construct points *D* through *H* and label them to match the figure in worksheet steps 7 and 8.
- At this point, have students identify which lines are parallel and which line is the transversal. Define *transversal* as a line that intersects, or crosses, two or more lines. Students can name the lines several ways, depending on which points they use. ***Which lines are parallel?*** [$\overleftrightarrow{DE} \parallel \overleftrightarrow{GB}$ or $\overleftrightarrow{DC} \parallel \overleftrightarrow{GA}$ or $\overleftrightarrow{CE} \parallel \overleftrightarrow{AB}$] ***Which line is the transversal?*** [$\overleftrightarrow{FC}$, $\overleftrightarrow{FA}$, $\overleftrightarrow{FH}$, $\overleftrightarrow{CA}$, $\overleftrightarrow{CH}$, or $\overleftrightarrow{AH}$.]

You can set the angle units and precision in **Edit | Preferences | Units.**

- Now model how to measure the size of an angle, such as $\angle CAB$. ***The points of the angle must be selected in the correct order, with the vertex as the second point.*** Identify the vertex as the point where the two sides of the angle meet. Select the three points, with point *A* second, and then choose **Measure | Angle.** Read the measure aloud. ***The* m *in front of the angle stands for "the measure of."***
- Model how to deselect all objects by clicking in blank space in the sketch before selecting each new angle to measure. You can also deselect all objects by pressing the Esc key one or more times.

3. If you want students to save their work, demonstrate choosing **File | Save As,** and let them know how to name and where to save their sketches.

DEVELOP

Expect students at computers to spend about 20 minutes.

4. Assign students to computers and distribute the worksheet. Tell students to work through step 18 and do the Explore More if they have time. Encourage students to ask their neighbors for help if they are having difficulty with the construction.

5. Let pairs work at their own pace. As you circulate, here are some things to notice.

   - In worksheet steps 7 and 8, be sure students label their figures the same as the illustration on the worksheet. It will be important for later discussions that all student figures are labeled the same. Remind students that they can choose the **Text** tool and double-click the label to change it.
   - In worksheet steps 9, notice whether students correctly name the vertex as the second point when naming angles with three points. ***What is the vertex of the angle you are measuring? How do you know?*** If needed, tell them that the vertex is the point where the two sides of the angle meet.
   - If the **Measure | Angle** command is not available, have students check that only the three points that define the angle are selected.
   - Some students may have difficulty finding all eight angles because they name one angle more than once. Suggest that they drag each measurement to place it near the vertex of the angle just measured.
   - In worksheet step 11, observe which students have trouble identifying the angle types. Discuss the terms with these students. ***Which angles are in the interior, or between, the parallel lines? Which angles are in the exterior, or outside, the parallel lines? Which angles are on alternate, or opposite, sides of the transversal? Which angles are on the same side of the transversal? What does* corresponding *mean?*** [In the same relationship or similar position]
   - When students are stating the relationship between angle pairs, some students may have difficulty identifying the relationship between same-side interior angles and same-side exterior angles. Have students

choose **Number | Calculate** and click on one angle measurement, the + key, the other angle measurement, and **OK** to find the sum of the angle measures.

- In worksheet steps 13–18, be sure that students understand what they are investigating. ***You just explored the relationship between the angles formed by the intersection of two parallel lines and a transversal. Suppose you start with the relationship between the angles. What can you say about the lines?***
- In worksheet step 17, if the angle degrees are given in tenths or hundredths, students may have difficulty moving the lines until they have two sets of four congruent angles. Have students get the measurements as close as possible and then make their conjectures.
- If students have time for the Explore More, they will construct two congruent alternate interior angles by marking $\angle CAB$, selecting point C as the center for rotation, and rotating $\overleftrightarrow{AC}$ by the marked angle. The resulting new line is parallel to $\overleftrightarrow{AB}$.

6. If students will save their work, remind them where to save it now.

SUMMARIZE

Project the sketch. Expect to spend about 15 minutes.

7. Gather the class. Students should have their worksheets with them. Open **Double Cross Present.gsp** and go to page "Parallel Lines." Begin the discussion by talking about the construction of the parallel lines and the transversal in worksheet steps 1–5. ***You constructed two parallel lines and a transversal. Can you identify some parallel lines in our classroom? Can you find a transversal?*** Students may suggest the following ideas.

The top and bottom of the window form parallel lines. A window side is a transversal.

The left and right sides of the door are parallel. The top of the door is a transversal.

The top and bottom edges of my book are parallel. The side of my book forms a transversal.

The shelves on the bookshelf are parallel. The left and right sides of the bookshelf are transversals.

How do you know the lines are parallel? [They are the same distance apart; they don't intersect.] Write the notation for showing that two lines are parallel on chart paper: $\overleftrightarrow{AB} \parallel \overleftrightarrow{CD}$. ***How do you know which lines are transversals?*** [A transversal is a line that intersects two or more lines.] Write the definitions for *parallel lines* and *transversal* on the chart paper.

8. In worksheet steps 11 and 12, name an angle type—corresponding angles, alternate interior angles, alternate exterior angles, same-side interior angles, and same-side exterior angles—and have a volunteer come up and identify the angle pairs and state their relationship. When discussing same-side interior and same-side exterior angles, review the term *supplementary angles* and what it means.

9. If needed, move the transversal so that it is not perpendicular to the parallel lines. Then discuss other types of angles found in the figure. Ask students to find acute and obtuse angles. ***What do you notice about the relationship among these angles?*** [The acute angles all have the same measure; the obtuse angles all have the same measure; the acute and obtuse angles are supplementary.]

10. Go to page "Converse." Discuss worksheet steps 13–18. ***Did you find that the converse of your conjectures in step 11 were true? Explain.*** Explain what the term *converse* means, especially if your students are unfamiliar with this idea. Students should have discovered that if two lines are intersected by a transversal and one of these statements is true, then the lines are parallel.

    - Corresponding angles are congruent.
    - Alternate interior angles are congruent.
    - Alternate exterior angles are congruent.
    - Same-side interior angles are supplementary.
    - Same-side exterior angles are supplementary.

11. ***If one of the angles is a right angle, what can you say about the transversal? How do you know?*** [If one angle is right, then all angles are right angles, and the transversal is perpendicular to the parallel lines.]

12. You may wish to have students respond individually in writing to this prompt. ***Suppose a transversal intersects two parallel lines and you know the measure of one angle, do you know the measure of all angles formed? Explain.*** [Yes, three other angles are the same measure and the remaining four angles are supplementary to the angle.]

13. If time permits, discuss the Explore More. Ask students to share their findings.

ANSWERS

10. There are two sets of four congruent angles.

11. Students should complete the chart as shown.

Angle Type	Pair 1	Pair 2	Relationship
Corresponding	∠*FCE* and ∠*CAB*	See answer 12.	Congruent
Alternate interior	∠*ECA* and ∠*CAG*	∠*DCA* and ∠*CAB*	Congruent
Alternate exterior	∠*FCE* and ∠*HAG*	∠*HAB* and ∠*DCF*	Congruent
Same-side interior	∠*ECA* and ∠*BAC*	∠*DCA* and ∠*GAC*	Supplementary
Same-side exterior	∠*FCD* and ∠*HAG*	∠*FCE* and ∠*HAB*	Supplementary

12. There are four pairs of corresponding angles. Those not listed are ∠*ECA* and ∠*BAH*, ∠*HAG* and ∠*ACD*, and ∠*GAC* and ∠*DCF*. Students should have recorded one of these pairs in the chart above.

18. The lines must be parallel.

24. Because the alternate interior angles were constructed to be congruent, the new line will be parallel to $\overleftrightarrow{AB}$.

Double Cross

Name:

In this activity you'll explore the relationships among the angles formed when parallel lines are intersected by a third line, called a *transversal.*

CONSTRUCT

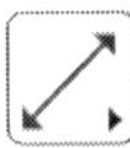

1. In a new sketch, construct $\overleftrightarrow{AB}$.

2. Construct point *C*, not on $\overleftrightarrow{AB}$.

3. Label the three points by clicking them in order from *A* through *C*.

4. Now you'll construct a line parallel to $\overleftrightarrow{AB}$ through point *C*. Select the line and the point and choose **Construct | Parallel Line.**

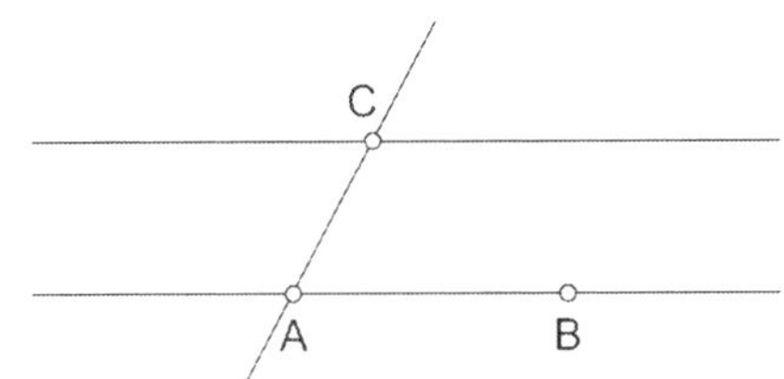

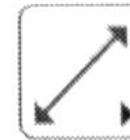

5. Construct $\overleftrightarrow{CA}$.

6. Drag points *C* and *A* to make sure the three lines are attached at those points.

7. Construct points *D, E, F, G,* and *H* as shown in the picture.

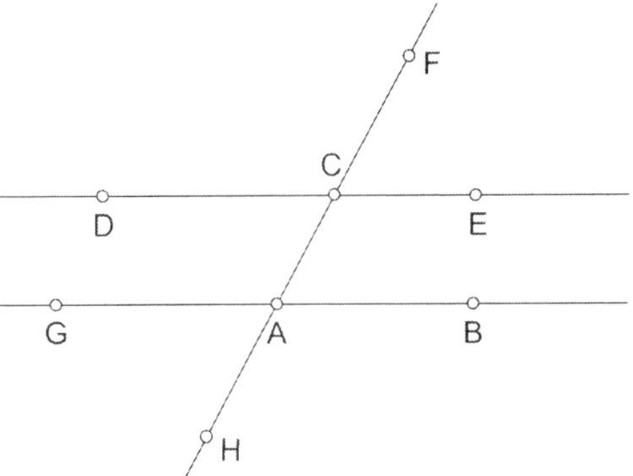

8. Label the points. You can double-click a label to change it.

EXPLORE

9. Now you'll measure the eight angles in your figure. Be systematic to make sure you don't measure the same angle twice. To measure an angle, select three points, with the vertex as your second point. Then choose **Measure | Angle.**

   Click in any blank space to deselect all objects after each measurement.

10. Drag point *A* or *B* and observe which angles stay congruent. Also drag the transversal $\overleftrightarrow{CA}$. (Be careful not to change the order of the points on your lines.) How many of the eight angles always appear to be congruent?

11. When two parallel lines are crossed by a transversal, the pairs of angles formed have specific names and properties. The chart shows one example of each type of angle pair. Fill in the chart with a second angle pair of each type, then explain how they are related.

Angle Type	Pair 1	Pair 2	Relationship
Corresponding	$\angle FCE$ and $\angle CAB$		
Alternate interior	$\angle ECA$ and $\angle CAG$		
Alternate exterior	$\angle FCE$ and $\angle HAG$		
Same-side interior	$\angle ECA$ and $\angle BAC$		
Same-side exterior	$\angle FCD$ and $\angle HAG$		

12. One of the angle types above has more than two pairs. Name that angle type and name the third and fourth pairs of angles of that type.

Angle Type	Pair 3	Pair 4	Relationship

13. Next, you'll investigate the converses of your conjectures. In a new sketch, draw two lines that are not quite parallel. Then construct a transversal.

14. Add points as needed.

15. Label the points *A* through *H* as shown in the picture.

16. Measure all eight angles formed by the three lines.

17. Move the lines until you have two sets of four congruent angles.

18. If two lines are crossed by a transversal so that corresponding angles, alternate interior angles, and alternate exterior angles are congruent, what can you say about the lines?

EXPLORE MORE

19. You can use the converse of the parallel lines conjecture to construct parallel lines. Construct a pair of intersecting lines $\overleftrightarrow{AB}$ and $\overleftrightarrow{AC}$ as shown.

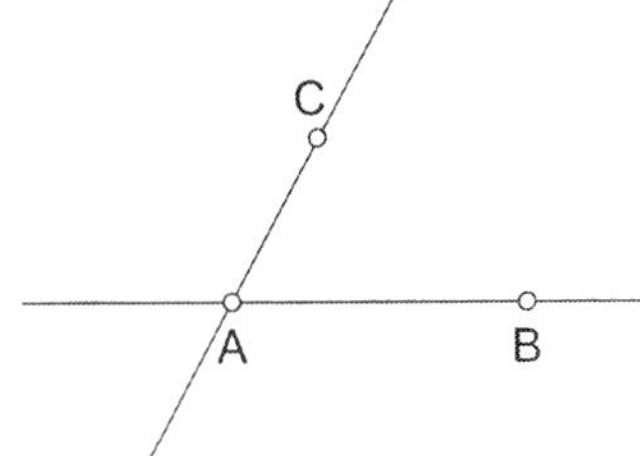

20. Label the points *A* through *C*.

21. Select, in order, points *C*, *A*, and *B* and choose **Transform | Mark Angle.**

22. Double-click point *C* to mark it as a center for rotation.

23. Select $\overleftrightarrow{AC}$. Then choose **Transform | Rotate** and click **Rotate.**

24. Drag the points of your sketch. Do the lines stay parallel? Explain why this method works.

Amazing Angles: Finding Transversal Angle Pairs

INTRODUCE

Project the sketch for viewing by the class. Expect to spend about 5 minutes.

1. Open **Amazing Angles.gsp** and go to page "Maze."
2. Explain, ***Today you'll be looking at the special angles that are formed when a transversal line intersects a pair of parallel lines. There are a surprising number of related angles that are formed, as you can probably see just by looking at the sketch. There are so many, in fact, that sometimes it's hard to keep them straight! You'll get a chance to identify all the different pairs of special angles as you work on different a-maze-ing challenges. Before you begin, I'll demonstrate how the sketch works.***
3. Ask students to identify which angle is congruent to $\angle BAC$ and help them choose the correct button (either *Show Vertical Angle, Show Corresponding Angle,* or *Show Alternate Exterior*) to show that angle. Depending on which button has been chosen, a new angle will appear on the sketch. Ask students, ***Now how could we easily return to our original*** $\angle$ ***BAC?*** Students should be able to see that you can simply click the same button again so that if $\angle BAC$ is corresponding to $\angle BAD$, then $\angle BAD$ will be corresponding to $\angle BAC$. Make sure to try all three angles that are congruent to $\angle BAC$.
4. Ask students, ***Now which angles are not congruent to*** $\angle$ ***BAC?*** Students might point to the appropriate angle, but encourage them to figure out which button you would have to press (*Show Supplementary Angle*) to express the correct relationship. Point out that the *Supplementary Angle* button shows the supplementary angle along one of the parallel lines, but there's also a supplementary angle along the transversal.

DEVELOP

Expect students at computers to spend about 15 minutes.

5. Assign students to computers and tell them where to locate **Amazing Angles.gsp.** Distribute the worksheet. Tell students to work through step 6 and do the Explore More if they have time.
6. Let pairs work at their own pace. As you circulate, here are some things to notice.
   - Help students use the correct mathematical notation for describing angles (use $\angle BAC$ instead of $\angle A$).
   - If students are having trouble distinguishing between alternate interior and alternate exterior angles, use the *Show Interior* and *Show*

Exterior buttons, which shade the regions between the two parallels and outside the two parallels, respectively.

- For worksheet steps 2 and 3, the last cell in the second column does not need to be filled out. Cells in the second column contain the button students pressed to get from the adjacent cell to the first cell in the following row.

SUMMARIZE

Project the sketch. Expect to spend about 10 minutes.

7. Bring students together and discuss their results for worksheet steps 2 and 3. Ask students to list the different paths they found. Challenge students to come up with paths that use the least and the most number of different buttons.
8. Drag point *E* so that the transversal is slanted in the opposite direction. Ask students how their answers on the worksheet would change using this configuration.
9. Drag point *D* so that the parallel lines are no longer horizontal. Ask students how their answers on the worksheet would change using this configuration.
10. Challenge students to figure out how they could change the configuration of lines (not shown on the sketch) so that the relationships between pairs of angles break down.

EXTEND

1. Challenge students to add a transversal that would be parallel to transversal $\overleftrightarrow{GB}$ and to identify all the pairs of related angles. Challenge students to add another line parallel to the existing pair of parallel lines and to identify the pairs of related angles.
2. ***What other questions might you ask about angles?*** Encourage all inquiry. Here are some ideas students might suggest.

   What are names for noncorresponding angles on the same side of the transversal? How are they related?

   How do you know that these angles that you say are congruent are actually congruent?

   How many ways are there of visiting all 8 angles without repetition?

ANSWERS

1. Answers will vary.
2. Again, answers will vary. Here's one possibility: $\angle BAC$ (vertical angle), $\angle BAD$ (supplementary angle), $\angle EAC$ (vertical angle), $\angle GEF$ (corresponding angle), $\angle HEA$ vertical angle), $\angle AEF$ (supplementary angle), $\angle GEH$ (vertical angle)
3. Answer will vary. Here's one possibility: $\angle AEF$ to $\angle HEA$, to $\angle EAC$, to $\angle BAD$
4. Answer will vary. Here's one possibility: $\angle AEH$ (corresponding) to $\angle AEF$ (supplementary)
5. Answers will vary. Here's one possibility: $\angle DAE$ to $\angle EAC$ (supplementary), to $\angle AEH$ (alternate interior), to $\angle AEF$ (supplementary)
6. Answer will vary.
7. Answers will vary. $\angle DAB$ is congruent to $\angle HEA$ when the lines *DC* and *HF* are parallel. $\angle DAB$ is congruent to $\angle AEF$ when the lines *DC* and *HF* are parallel and both are perpendicular to line *BG* (in which case they are both right angles), or when the two blue lines intersect to form the congruent sides of an isosceles triangle.

Amazing Angles

Name:

In this activity you'll be solving several problems that will require you to find relationships between the eight angles formed at the intersections of two parallel lines with a transversal.

EXPLORE

1. Open **Amazing Angles.gsp** and go to page "Maze." Currently, $\angle BAC$ is marked. Use the buttons to visit every one of the eight angles created by the transversal line. Record the angles you've visited as you go. You may visit some angles more than once.

2. Start at $\angle DAE$ (you may have to click a few buttons to get there). Visit every one of the eight angles *only once*. Record the order of the angles and the buttons you pressed.

Angle	Button
$\angle DAE$	

3. Start at $\angle AEF$. How can you get to $\angle BAD$ in exactly three steps?

Angle	Button
$\angle AEF$	
$\angle BAD$	

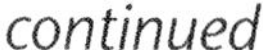

4. Start at ∠*BAD*. Can you get to ∠*AEF* without visiting ∠*BAC* or ∠*CAE*? If so, record your steps.

5. Describe how you can visit all the interior angles without stepping in the exterior region. Start at ∠*DAE*. Try to find more than one way.

6. Create your own "angle maze" problem and write a solution.

EXPLORE MORE

7. Go to page "Explore More." Here there is a red transversal that cuts two nonparallel lines. Predict which pairs of angles are either congruent or supplementary. Drag the points on the sketch to change the relationship between lines *DC* and *HF* and drag points *A* and *E* to change the slant of the transversal. Explain the conditions under which ∠*DAB* is congruent to ∠*HEA*. Can ∠*DAB* ever be congruent to ∠*AEF*?

2

Triangle Properties

Triangle Totals: Angles in a Triangle

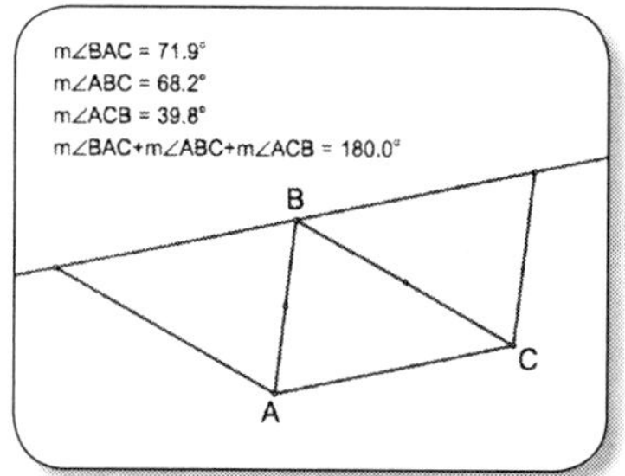

Students measure the angles of a triangle to write a conjecture about the sum of the angle measures in a triangle. Then they add an auxiliary line and rotate the triangle around the midpoints of two sides to develop a proof of the conjecture.

Meet the Isosceles Triangle: Properties of Isosceles Triangles

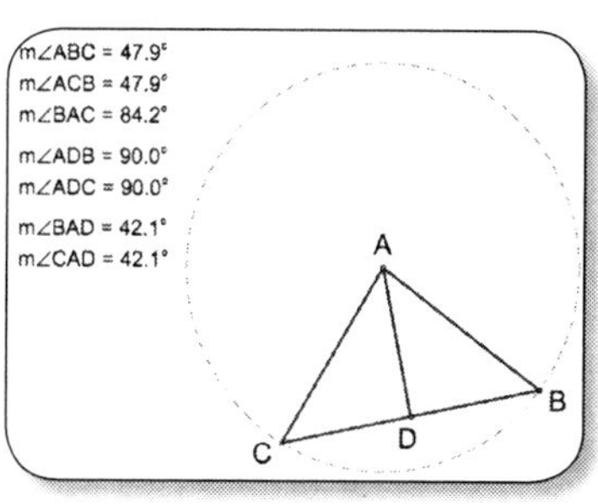

Students construct an isosceles triangle, measure the angles, and observe relationships among them. Then they construct the median of the base and the vertex angle, measure the angles formed, and discover that the median to the vertex angle is the perpendicular bisector of the base and bisects the vertex angle.

Triangle Pretenders: Classifying Triangles

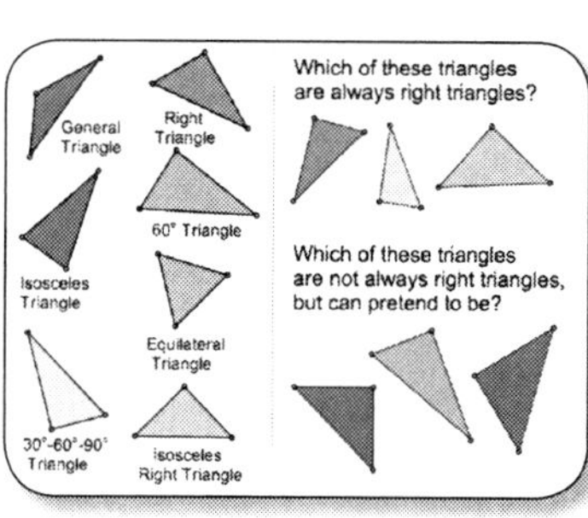

Students drag vertices of various Sketchpad triangles to discover which are constructed to have specific characteristics. As they make distinctions on the basis of characteristics, they understand deeply the definitions of various classifications of triangles, their properties, and the relationships among them.

Three Pairs: Triangle Congruence Properties

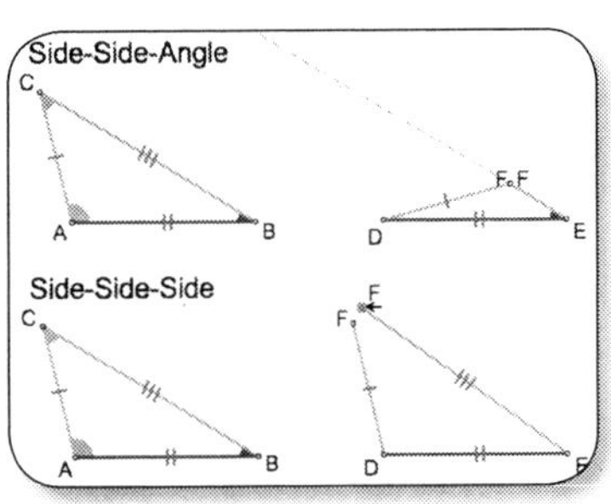

Students use a prepared sketch to explore possible triangle congruence conditions. They are given different combinations of three congruent parts (such as three congruent sides) and asked whether those parts are sufficient to guarantee that the triangles are congruent.

Tiling with Triangles: Sums of Interior Angles in Polygons

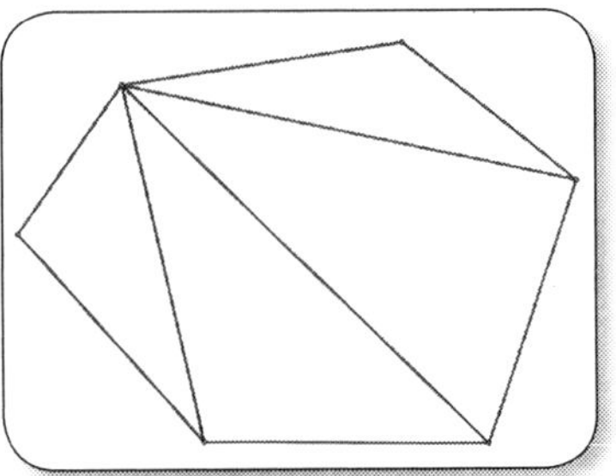

As students divide a polygon's area into triangles, they discover a relationship between the number of triangles that make up the area and the sum of the measures of the interior angles in a polygon. This geometric relationship is the starting point for deriving an algebraic expression.

Small, Medium, and Large: Triangle Inequalities

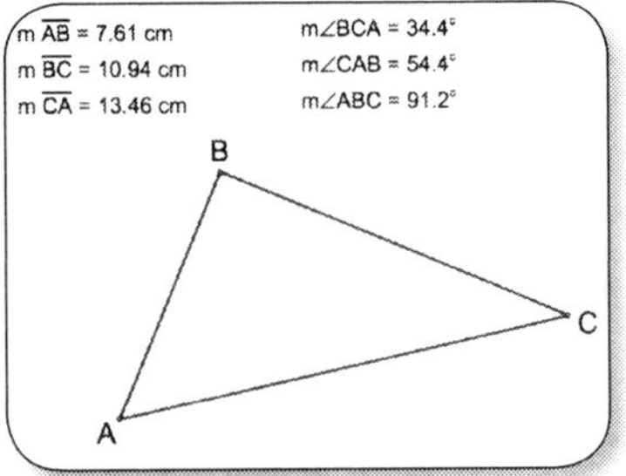

Students construct a triangle and make observations about side lengths. They notice that the sum of two side lengths is always greater than the third side length, and that the longest side is opposite the largest angle and the shortest side is opposite the smallest angle.

Tilted Squares: Finding Square Roots

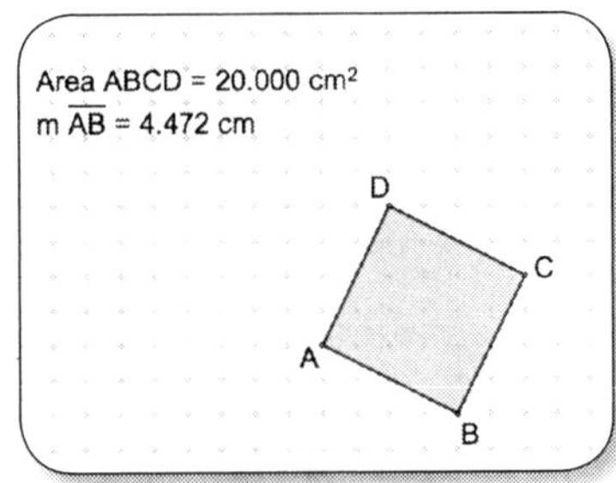

Students use Sketchpad as a geoboard to explore the relationship between squares and square roots. They measure the areas of different-sized squares and find the square root of each area by measuring a side length of the square. They make a table of their measurements, creating a list of squares and square roots.

Squaring the Sides: The Pythagorean Theorem

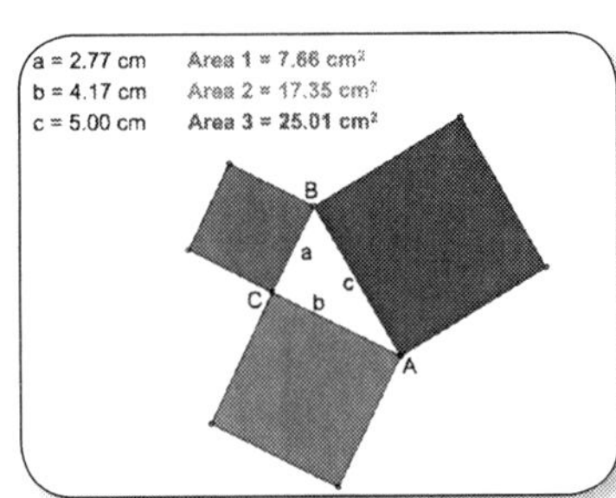

Students investigate the Pythagorean Theorem by constructing squares on the sides of a right triangle and exploring the relationship among their areas. Students discover that for any right triangle, the sum of the areas of the squares on the legs is equal to the area of the square on the hypotenuse.

Triangle Totals: Angles in a Triangle

INTRODUCE

Project the sketch for viewing by the class. Expect to spend about 10 minutes.

1. Open Sketchpad and enlarge the document window so it fills most of the screen. As you demonstrate, make lines thick and labels large for visibility.

2. Explain, ***Today you're going to use Sketchpad to make a conjecture about the sum of the measures of the angles of a triangle. What do you know about triangles already?*** Students may make the following responses.

   A triangle has three sides.

   It's a polygon.

   A triangle has three angles.

   Some of your students may already know that the sum of the angle measures in any triangle is 180°. Tell them that in this activity, they will prove that it is true.

3. ***I'll show you how to construct a triangle and measure an angle using Sketchpad.*** Model worksheet steps 1–6. Here are some tips.

   - First show students how to change the Sketchpad preferences so the angle measures are in tenths of a degree. Choose **Edit | Preferences | Units** and set Angle Precision to **tenths.** Be sure Angle Units are set to **degrees.** Then click OK.
   - Demonstrate how to construct a triangle. Students can also construct a triangle by selecting the points and choosing **Construct | Segments.**
   - ***The three points where the sides of the triangle meet are called vertices. We'll label the three vertices A, B, and C.*** Model how to use the **Text** tool to click on each vertex to label it. Tell students they can double-click on a vertex to edit a label.
   - Review how to name a triangle by its vertices. ***This is triangle ABC.*** Show students how to write $\triangle ABC$ using the triangle symbol.
   - Go over how to name an angle. Point to angle *A*. ***What is the name of this angle?*** [Angle *BAC*, angle *CAB*, or angle *A*] ***When naming an angle using three points, always name the vertex as the second point.***
   - Demonstrate how to measure an angle by selecting three points with the vertex as the second point. Then choose **Measure | Angle.** Have a

volunteer read the angle measure. Review that "$m\angle$" means "measure of angle," if necessary.

- ***You will use the Sketchpad Calculator to find the sum of the angle measures.*** Show students how to open the Sketchpad Calculator and click on the measurement to enter it into a calculation.
- Explain that if students make an error during their construction, they can choose **Edit | Undo** to undo the most recently performed action.

4. If you want students to save their work, demonstrate choosing **File | Save As,** and let them know how to name and where to save their files.

DEVELOP

Expect students at computers to spend about 25 minutes.

5. Assign students to computers and distribute the worksheet. Tell students to work through step 16 and do the Explore More if they have time. Encourage students to ask their neighbors for help if they are having difficulty with the construction.
6. Let pairs work at their own pace. As you circulate, here are some things to notice.
   - In worksheet step 5, watch for students who measure the same angle twice. ***Show me the angles you measured. How can you be sure that you didn't measure the same angle twice?*** Be sure students understand that the vertex of the angle must be the second point selected. ***What's the vertex of this angle?***
   - In worksheet step 7, when students drag a vertex, have them make a variety of different types of triangles. Ask students to make obtuse, right, acute, scalene, isosceles, and equilateral triangles. ***What happens to the measure of each angle? What happens to the angle sum?***
   - In worksheet step 8, if students quickly make a conjecture, ask them to try to disprove their conjecture by dragging the vertices to make a triangle in which the sum of the angle measures does not equal 180°. ***Can you construct a triangle in which the sum of the angle measures is not 180 degrees?***
   - In worksheet step 9, check that students understand what parallel lines are. ***What are parallel lines?*** [Two lines in the same plane that never intersect]

- In worksheet step 12, explain that the point will briefly flash after it is marked. Otherwise, the sketch will appear the same. Be sure students enter 180 for degrees.
- In worksheet step 15, as students are dragging point *B*, check that they understand the relationship among the angles formed when two parallel lines are intersected by a transversal. ***What is a transversal?*** [A line that intersects two or more lines] ***What do you know about the angles formed when a transversal intersects two parallel lines?*** For this activity students will need to know that alternate interior angles are congruent.
- In worksheet step 16, have students identify the straight angle and its measure. Point to the three angles around point *B*. ***What type of angle is formed by these three angles?*** [Straight angle] ***What is its measure?*** [180°] ***How does this help you prove your conjecture?***
- If students have time for the Explore More, they will make a conjecture about the sum of the angle measures of quadrilaterals.

7. If students will save their work, remind them where to save it now.

SUMMARIZE

Project the sketch. Expect to spend about 10 minutes.

8. Gather the class. Open **Triangle Totals Present.gsp.** Students should have their worksheets with them. ***What conjecture did you make about the sum of the measures of the angles of a triangle? Are any sums possible? Does it hold true for all types of triangles?*** Students may make the following response: *We made all sorts of triangles and found that no matter what type of triangle we constructed, the sum of the measures of the angles always equaled 180 degrees.*
9. ***How did you prove that your conjecture is true?*** Let volunteers explain their thinking, and then walk through the proof with students. Press *Show Why* and then the new buttons that appear. ***The line containing point B is parallel to what?*** [Segment *AC*] ***What is segment AB then?*** [A transversal] ***What is segment CB?*** [A transversal]
10. ***What type of angle do the three angles around point B form?*** [Straight angle] ***What is the measure of a straight angle?*** [180°] ***If segment BA is a transversal, then which angle in the triangle ABC is congruent to***

this angle? Point to the left angle that has a vertex on point *B*. [Angle *BAC*] ***How do you know the angles are congruent?*** [When a transversal intersects two parallel lines, alternate interior angles are congruent.]

11. ***If segment BC is a transversal, then which angle in triangle ABC is congruent to this angle?*** Now point to the right angle that has a vertex on point *B*. ***Why are the angles congruent?*** [When a transversal intersects two parallel lines, alternate interior angles are congruent.]

12. The middle angle with a vertex on point *B* is already an angle of the triangle. Students should see that the angles that formed the straight angle are the three angles of the triangle. ***Because the measure of a straight angle is 180 degrees, then the sum of the measures of the angles of any triangle is 180 degrees.*** Explain that this is known as the Triangle Sum Theorem.

13. If time permits, discuss the Explore More. ***What is your conjecture? Did it hold true for all types of quadrilaterals?*** Let volunteers explain their thinking. Students should discover that the sum of the angle measures of any quadrilateral is 360°.

14. You may wish to have students respond individually in writing to this prompt. ***Suppose a triangle has two angles that measure 45 degrees each. What is the measure of the third angle? Explain your reasoning.*** [90°; $180 - (45 + 45) = 90$]

EXTEND

1. ***Can an angle in a triangle be any measure?*** [No, the sum of the three angles must equal 180°.] ***What do you think is the greatest angle measure for an angle in a triangle? What is the least angle measure?*** [The angle measures must be between 0° and 180°, not including those measures.]

2. ***Can a triangle have two right angles? Explain.*** [No, then the third angle would have a measure of 0°, which is not possible.]

ANSWERS

8. The sum of the measures of the angles in any triangle is 180°.

16. Students need to show that the three angles that form a straight angle at point *B* are congruent to the three angles of the original triangle.

Because the horizontal line was constructed parallel to the opposite side of the triangle, the angles presented here marked with a single arc are congruent because they are alternate interior angles. The same argument holds for the angles marked with two arcs. The third unmarked angle at point *B* is the remaining angle of the original triangle. So the three angles, which form a straight angle at point *B* (and whose measures thus add up to 180°), are each congruent to the three angles of the original triangle. Therefore, the three angle measures of the triangle must also add up to 180°.

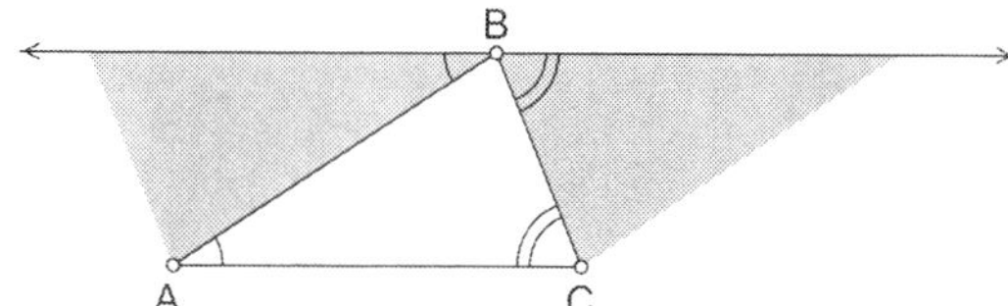

17. Answers will vary.
18. The sum of the measures of the angles in any quadrilateral is 360°.
19. A quadrilateral can be split into two triangles, so the sum of the measures of the angles is twice 180°.
20. For a polygon with n sides, the sum of the measures of the angles is $180°(n - 2)$.

Triangle Totals

Name:

In this activity you'll investigate and make a conjecture about the sum of the measures of the angles in a triangle. Then you'll try to show that your conjecture is true.

CONSTRUCT

1. First you'll change the Sketchpad preferences so angle measures are in tenths of a degree.

   In a new sketch, choose **Edit | Preferences** and go to the Units panel.

   Set Angle Units to **degrees** and Angle Precision to **tenths.**

   Click **OK.**

2. Construct three points.

m∠CAB = 43.2°
m∠ABC = 87.6°
m∠BCA = 49.1°

B
A
C

3. Label the points by clicking on each one.

4. Construct △*ABC*.

5. Now you'll measure the three interior angles.

   To measure an angle, select three points with the vertex as your second point. Then choose **Measure | Angle.**

6. Next you'll calculate the sum of the angle measures.

   Choose **Number | Calculate** to show Sketchpad's Calculator.

   Click once on a measurement to insert it into a calculation.

   Use the + on the Calculator keypad to find sums.

   When you are done, click **OK.**

EXPLORE

7. Drag a vertex of the triangle and observe the angle sum.

8. What is the sum of the measures of the angles in any triangle?

Follow these steps to explore why your conjecture is true.

CONSTRUCT

9. Select point *B* and $\overline{AC}$ and choose **Construct | Parallel Line.**

10. Select $\overline{AB}$ and $\overline{CB}$ and choose **Construct | Midpoint.**

11. Select the three vertices of $\triangle ABC$ and choose **Construct | Triangle Interior.**

12. Now you'll rotate the triangle interior by 180° about a midpoint.

    To mark one of the midpoints as a center for rotation, double-click the point.

    Then select the triangle interior and choose **Transform | Rotate.** In the Rotate dialog box, enter 180 for degrees and click **Rotate.**

13. Give the new triangle interior a different color by selecting it and choosing **Display | Color.**

14. Mark the other midpoint as a center and rotate the original triangle interior by 180° about this point. Give this new triangle interior a different color.

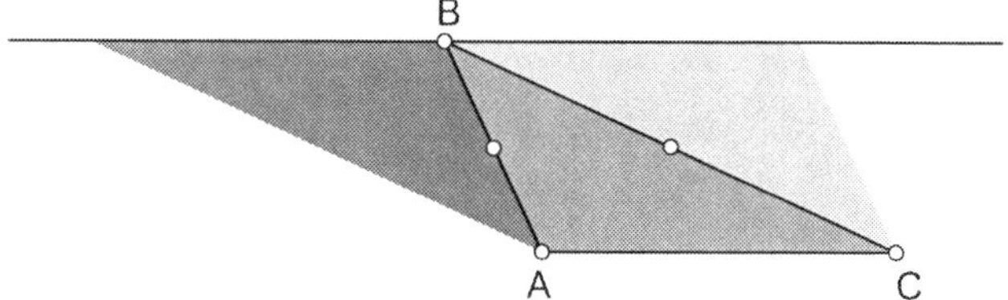

EXPLORE

15. Drag point *B* and observe how the three triangles are related to each other and to the parallel line.

16. Explain how each of the three angles at point *B* is related to one of the three angles in the original triangle. Explain how this demonstrates your conjecture from step 8.

EXPLORE MORE

17. Make a prediction about the sum of the measures of the angles of a quadrilateral.

18. Construct a quadrilateral. Measure each interior angle and find the sum of the measures. Drag a vertex of the quadrilateral and observe the angle sum. What is the sum of the measures of the angles in any quadrilateral?

19. Explain why your conjecture from step 18 is true. (*Hint:* Construct a diagonal of the quadrilateral.)

20. What is the sum of the measures of the angles in *any* polygon?

Meet the Isosceles Triangle: Properties of Isosceles Triangles

INTRODUCE

Project the sketch for viewing by the class. Expect to spend about 5 minutes.

1. Open Sketchpad and enlarge the document window so it fills most of the screen. As you demonstrate, make lines thick and labels large for visibility.
2. Explain, ***Today you're going to use Sketchpad to construct an isosceles triangle and investigate its properties. What is an* isosceles triangle*?*** Work with students to come up with a common definition. Here is a sample definition: An isosceles triangle is a triangle that has at least two sides of equal length. Draw an isosceles triangle on chart paper. Make single tick marks on the two sides that are congruent. Explain that the tick marks mean the two sides are of equal length, or are *congruent.*
3. Label and define the following parts of the isosceles triangle: *leg, vertex angle, base,* and *base angles.* ***The two sides that have equal lengths are called* legs. *The third side is called the* base. *The two angles opposite the legs are called* base angles. *The third angle opposite the base is called the* vertex angle.** Do not tell students that the base angles have the same measure; they will discover this property on their own.
4. ***Now I'll demonstrate how to construct an isosceles triangle using Sketchpad.*** Model the triangle construction in worksheet steps 1–7. Here are some tips.
   - In worksheet steps 1 and 2, construct a circle using the **Compass** tool, and then label the center *A* and the radius point *B* using the **Text** tool. ***Point B is called a* radius point. *Why do you think it's called that? What is a* radius *of a circle?*** [A line segment with one endpoint at the center and the other endpoint on the circle] Help students understand that any point on the circle is a radius point, because you can construct a radius of the circle from that point to the center of the circle.
   - In worksheet steps 3 and 4, model how to construct the two radii. Instead of using the **Segment** tool, students can also select the center and radius point and choose **Construct | Segment.**
   - In worksheet step 6, stress the importance of making sure that $\overline{AC}$ is actually a radius. Explain that students can choose **Edit | Undo** to undo the most recent step, if needed.

- In worksheet step 7, finish the third side of the triangle by constructing $\overline{BC}$. ***The triangle is done. Now you'll construct the triangle and explain why it's always an isosceles triangle. Then you'll explore some of the properties of isosceles triangles.***

5. If you want students to save their work, demonstrate choosing **File | Save As,** and let them know how to name and where to save their files.

DEVELOP

Expect students at computers to spend about 15 minutes.

6. Assign students to computers. Distribute the worksheet. Tell students to work through step 20 and do the Explore More if they have time. Encourage students to ask their neighbors for help if they are having difficulty with the construction.

7. Let pairs work at their own pace. As you circulate, here are some things to notice.

   - In worksheet step 8, ask students to share what they observe as they drag the vertices of the triangle. ***What happens to the radii, $\overline{AB}$ and $\overline{AC}$, when you drag the vertices?*** Help students see that the radii get shorter or longer as the circle increases or decreases in size, but they always remain equal to each other in length.

   - In worksheet step 11, be sure students name the vertex as the second point when measuring the interior angles of the triangle. Also, watch for students who measure the same angle twice. ***Name the angles you have measured. Are angle ABC and angle CBA the same or different angles? How do you know?*** [The same angle; the vertex of both angles is point *B*.]

   - In worksheet step 12, the angle measurements will be in hundredths of a degree unless you have students change the Sketchpad preferences before they measure the angles. Students can choose **Edit | Preferences** and in the Units panel, set Angle Precision to **tenths** or **units.** Have students read an angle measurement to verify that they understand what it means. ***What is the measure of angle ABC?*** Explain that the notation "$m\angle$" means "measure of angle."

   - In worksheet steps 14–16, students will construct a median of the triangle. Review the definition of *median.* ***A median of a triangle is a line segment from a vertex to the midpoint of the opposite side.***

- In worksheet step 19, students should notice that $\angle ADB$ and $\angle ADC$ are always 90°, which means $\overline{AD}$ is perpendicular to the base, $\overline{CB}$. ***What can you tell me about the relationship between*** $\overline{AD}$ ***and*** $\overline{CB}$***?*** [They are perpendicular because they form 90° angles.] Remind students that when they constructed point *D*, they made it the midpoint of the base. ***What do you know about the lengths of*** $\overline{CD}$ ***and*** $\overline{BD}$***?*** [They are equal in length because point *D* is the midpoint.] ***What does*** $\overline{AD}$ ***do to*** $\overline{CB}$***?*** [It bisects it.] ***What does that make*** $\overline{AD}$***?*** [A perpendicular bisector]
- Students should also observe that $\angle BAD$ and $\angle CAD$ are always congruent, which means $\overline{AD}$ is an angle bisector. ***What can you tell me about the relationship between*** $\overline{AD}$ ***and*** $\angle CAB$***?*** [Segment *AD* bisects $\angle CAB$ because $\angle BAD$ and $\angle CAD$ are congruent.]
- If students have time for the Explore More, they will use the properties of an isosceles triangle to construct isosceles triangles, using different methods. Be sure students clearly describe their construction methods and explain what properties they used. The more time you give students, the more methods they will come up with. Give students hints to start their thinking, if necessary. Have students drag vertices of their figures to make sure their constructions are correct. Isosceles triangles that fall apart and can turn into other shapes are underconstrained. A construction that stays an isosceles triangle but that can't take on all possible shapes of an isosceles triangle is overconstrained.

8. If students will save their work, remind them where to save it now.

SUMMARIZE

Project the sketch. Expect to spend about 10 minutes.

9. Gather the class. Students should have their worksheets with them. Open **Meet Isosceles Tri Present.gsp.** Go to page "Angles." Begin the discussion by having students identify the different parts of the isosceles triangle. ***Which are the legs? How do you know?*** [Sides *AC* and *AB* are the legs because they are the same length.] ***How do you know they are the same length?*** [They are both radii of the circle. All radii of the same circle are the same length.] ***Which side is the base?*** [Side *CB*] ***Which are the base angles?*** [Angles *ACB* and *ABC*] ***Which is the vertex angle?*** [Angle *CAB*]

10. ***Today you explored some properties of isosceles triangles.*** Start a chart listing the properties of isosceles triangles. ***What do you know about the sides of isosceles triangles?*** Write that at least two sides are congruent.

11. Now press *Animate* on the sketch. ***What did you observe about the angle measures?*** Students should reply that the base angles are always congruent. Add this property to the chart.

12. Now go to page "Median." Have students identify the median and define it. ***You constructed a median of the base and the vertex angle. What did you discover about the median?*** Press *Animate*. Then fill in the chart with students' observations, clarifying their language as needed.

Properties of Isosceles Triangles
At least two sides are congruent.
Base angles are congruent.
The median of the base and the vertex angle is a perpendicular bisector of the base and an angle bisector of the vertex angle.

13. If time permits, discuss the Explore More. Have students review their construction methods and the properties they used in each.

14. ***If the vertex angle of an isosceles triangle is 90 degrees, what are the measures of the other two angles? Explain how you know.*** [The two base angles are equal and must add up to 90° because the sum of the measures of the angles in a triangle is 180° and 180 − 90 = 90. Therefore, the base angles must both measure 45°.]

EXTEND

Have students discuss the following questions with a partner. Then discuss as a whole class.

1. Can an isosceles triangle have an obtuse angle (measure greater than 90°)? Explain. [Yes, the vertex angle can be obtuse.]

2. Can the base angles of an isosceles triangle be right angles? Explain. [No. If they measured 90°, then their sum would be 180°. It would not form a triangle because the sum of the three angles of a triangle must equal 180°.]

3. Can an isosceles triangle be an equilateral triangle (all sides equal in length)? [Yes, at least two sides must have equal length.]

ANSWERS

9. Two sides of the triangle are radii of the same circle. This guarantees that two sides of the triangle are always congruent. Therefore, the triangle is always isosceles.

13. The base angles of an isosceles triangle have equal measure. The vertex angle usually has a different measure.

20. The median of the base and vertex angle of an isosceles triangle is the perpendicular bisector of the base and the angle bisector of the vertex angle.

21. Students may find one or more of the following methods.

Method: Construct a line *AB* and a point *C* not on the line. Reflect point *C* across the line. Triangle *ACC′* is isosceles.
Property: An isosceles triangle has reflection symmetry.

Method: Construct segment *AB* and its midpoint *C*. Construct a perpendicular through point *C*. Construct point *D* on the perpendicular. Triangle *ABD* is isosceles.
Property: The perpendicular bisector of the base of an isosceles triangle passes through the vertex of the vertex angle.

Method: Construct ray *AB* and ray *AC*. Construct the angle bisector of $\angle BAC$. Construct point *D* on the angle bisector and a line through point *D* perpendicular to the angle bisector. Construct the points of intersection *E* and *F* of this line with ray *AB* and ray *AC*. Triangle *AEF* is isosceles.
Property: The angle bisector of the vertex angle of an isosceles triangle is perpendicular to the base.

Method: Construct ray *BA* and ray *AC*. Select points *C*, *A*, and *B* and, choose **Transform | Mark Angle**. Mark point *B* as a center. Rotate ray *BA* by the marked angle. Construct the point of intersection, *D*, of this ray and ray *AC*. Triangle *ABD* is isosceles.
Property: The base angles of an isosceles triangle are equal in measure.

Meet the Isosceles Triangle

Name:

In this activity you'll learn how to construct an *isosceles triangle* (a triangle with at least two sides the same length). Then you'll discover properties of isosceles triangles.

CONSTRUCT

1. In a new sketch, construct a circle.

2. Label the center *A* and the radius point *B*.

3. Construct radius $\overline{AB}$.

4. Construct another radius.

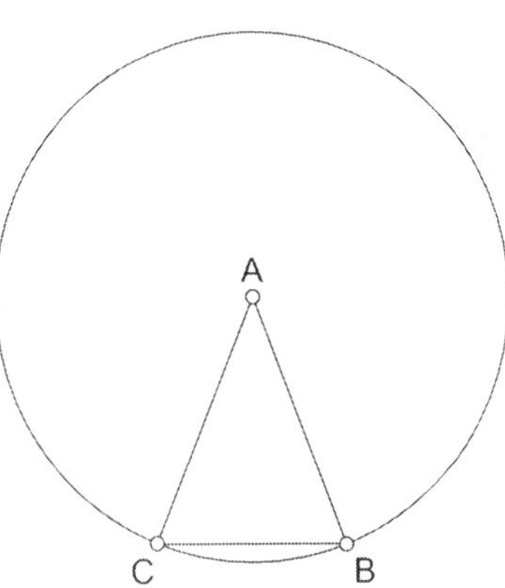

5. Label the new radius point *C*.

6. Drag point *C* to make sure radius $\overline{AC}$ is attached to the circle.

7. Construct $\overline{BC}$.

EXPLORE

8. Drag each vertex of your triangle to see how it behaves.

9. Explain why the triangle is always an isosceles triangle.

Now you'll explore some properties of isosceles triangles.

10. Hide the circle by selecting it and choosing **Display | Hide Circle.**

11. Measure the three interior angles in the triangle.

    To measure an angle, select three points with the vertex as your second point. Then choose **Measure | Angle.**

12. Drag the vertices of your triangle and observe the angle measures.

13. Angles *ACB* and *ABC* are the base angles of the isosceles triangle. Angle *CAB* is the vertex angle. What do you observe about the angle measures?

CONSTRUCT

14. Select $\overline{BC}$ and choose **Construct | Midpoint.**

15. Label the midpoint *D*.

16. Construct $\overline{AD}$, the median between the base and vertex angle.

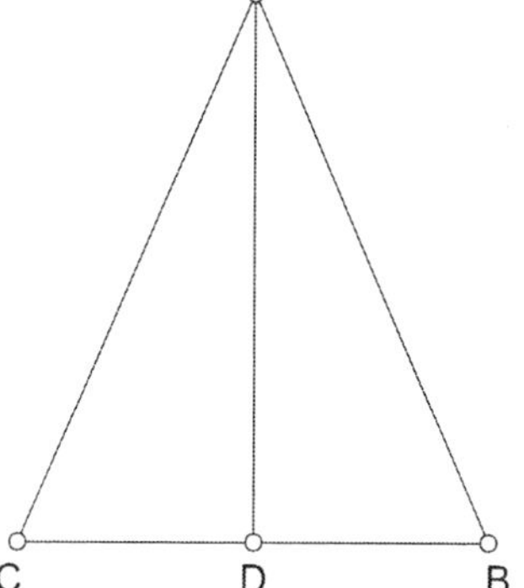

EXPLORE

17. Measure $\angle ADB$ and $\angle ADC$.

18. Measure $\angle BAD$ and $\angle CAD$.

19. Drag the vertices of your triangle and observe the new angle measures.

20. Segment *AD* is the median of the base and the vertex angle. What do you observe about the median of the base and the vertex angle of an isosceles triangle?

EXPLORE MORE

21. How many different ways can you construct an isosceles triangle? Try methods that use just the freehand tools on the left and methods that use the Construct and Transform menus. Write a brief description of each construction method along with the properties of isosceles triangles that make that method work.

Triangle Pretenders: Classifying Triangles  ACTIVITY NOTES

INTRODUCE

Project the sketch for viewing by the class. Expect to spend about 10 minutes.

1. Open **Triangle Pretenders Present.gsp** and go to page "Drag Test." You might press *Show Distances* to show that all the triangles are isosceles.

2. Say, ***Some of these Sketchpad shapes have been constructed to always be isosceles triangles, and some are just pretending to be.*** Drag vertex *A*, and then *B* and *C* of the blue triangle, and see that it is no longer isosceles. ***What stays the same as I drag? What changes? What can you say about the shape?*** [It's always a right triangle] Students may offer that it's a general triangle, or scalene triangle. Suggest that they be as specific as possible. Help clarify that it's not always an isosceles triangle (although it may pretend to be), but it's always a right triangle.

3. As you drag other parts of the blue triangle, ask, ***Does it stay a right triangle no matter what I drag?*** [Yes] ***The drag test showed that the right-triangle shape was pretending to be isosceles.*** Although the blue triangle initially looked like an isosceles right triangle, it was not constructed to always have two sides the same length. As you discuss isosceles triangles, remember not to say *equal sides* if you mean equal in length. Your curriculum may use the term *congruent* for line segments that are equal in measure.

4. Review how a polygon is named. ***One name for this Sketchpad shape is right triangle BCA. What is another way to name this triangle?*** Students might give any of the other five names: *ABC, BAC, ACB, CBA,* or *CAB.*

5. Refer to the purple triangle. ***This second shape doesn't have its vertices named.*** Show students how to click on a vertex with the **Text** tool to show its label. Once all the vertices are labeled, ask students to name the triangle by its vertices. For example, they might say *DEF, FED,* and so on. ***What can you say about the shape DEF?*** [It's a general triangle. It doesn't have more specific characteristics.]

6. ***What about shape XYZ? Is it always an isosceles triangle or is it a pretender?*** Drag vertex *X, and* then vertex *Y.* ***It still looks isosceles. What might this shape actually be?*** [It looks equilateral.] ***Let's try to drag the other vertex.*** When students see that it is no longer isosceles, ask them, ***What does stay the same on this shape?*** [One angle always stays the same size.] Select in order points *Y, X,* and *Z* and choose **Measure | Angle.** Then drag the vertices to verify that the angle always measures 60°. ***We'll call these triangles "60° triangles."***

7. ***What about shape PQR?*** Drag each vertex to show that triangle *PQR* is always an isosceles triangle. As students begin work on the activity, let them know that identifying the pretenders is only part of the task; they must also describe as specifically as possible what each pretender really is. Encourage students to ask their classmates for help with Sketchpad if they are having difficulty.

DEVELOP

Expect students at computers to spend about 25 minutes.

8. Assign students to computers and tell them where to locate **Triangle Pretenders.gsp.** Distribute the worksheet. Tell students to work through step 6 and do the Explore More if they have time.

9. Let pairs work at their own pace. As you circulate, here are some things to notice.

   - For students who may be jumping to conclusions, ask, ***Have you tried to drag each vertex?***

   - Use questions to help students describe the pretenders completely.

     What stays the same as you drag this vertex or edge? What changes?

     How are these two sides related?

     What is happening to this angle (side)? To the other angles (sides)?

   - Encourage students to identify each shape as specifically as possible. For example, on page "Isosceles," once students have identified that the triangles were constructed to have two sides of equal measure, ask, ***Are there triangles that are isosceles but also have more specific names?*** [Equilateral triangles]

   - If your curriculum uses an exclusive definition for isosceles triangles (exactly two sides of equal measure), you might need to comment on the use of a different definition on page "Isosceles." There the inclusive definition (at least two sides of equal measure) includes the equilateral triangle as isosceles.

10. You might collect students' Explore More work on a flash drive or ask them to save their sketches where they can be displayed during the class discussion.

SUMMARIZE

Project the sketch. Expect to spend about 10 minutes.

11. Gather the class. Students should have their worksheets with them. You might invite selected pairs to demonstrate their work on page "Isosceles" or "Right." Ask demonstrators to explain why they made particular choices. Alternatively, you might lead a class discussion around the solutions provided in **Triangle Pretenders Present.gsp.**

12. If you have time, discuss the Explore More. Have students share their results—triangles that passed the drag test and the pretenders they made—and explain how they used the properties of the specific triangle to construct the shape in Sketchpad. You might spend another day with students, showing their construction work. These constructions could also extend to projects that are presented later.

13. ***What have you learned through this investigation?*** Help students articulate whatever they have learned. Include the objectives of the lesson. Students might offer suggestions such as these, or you might want to bring these up.

    - The drag test reveals pretenders—shapes that may *look* like a particular shape, but are not constructed to have the required characteristics.
    - A general triangle can pretend to be any of the more specific triangles.
    - Sometimes definitions of shapes are different. It is important to understand the definitions being used, particularly when you are building a hierarchy of shapes.
    - The equilateral triangle, though it is isosceles using the inclusive definition, can't pretend to be any other triangles.
    - The 60° triangle can pretend to be a right triangle (30°-60°-90°) or an equilateral triangle.

14. After exploring students' observations about relationships among triangles, you might suggest that students create a tree diagram or Venn diagram showing those relationships. They might draw diagrams in Sketchpad, like the ones on pages "Venn Diagram" or "Tree Diagram" in **Triangle Pretenders Present.gsp.** Encourage discussion about where to put 60° triangles and how to place the subset of 30°-60°-90° triangles.

EXTEND

What other questions might you ask about relationships among shapes? Here are some ideas students might suggest.

Are the pretenders really pretending? Aren't they actually the shape they're pretending to be, even if only for that instant?

Are there other special kinds of triangles?

Is there a name for an isosceles triangle that is not equilateral?

Is an isosceles triangle with one angle of 60° necessarily equilateral?

If shape A can pretend to be shape B, are all B-shapes a subset of A-shapes?

Can you classify shapes other than triangles? What classifications are used?

ANSWERS

1. *JKL:* general triangle; *NOP:* 60° triangle; *DEF:* equilateral triangle; *RST:* isosceles triangle
2. *ABC:* right triangle; *EFG:* isosceles triangle; *MNO:* isosceles right triangle; *RUS:* general triangle; *HIT:* 60° triangle
3. *BIG:* isosceles right triangle; *CUT:* right triangle; *FLY:* isosceles triangle; *HEN:* general triangle; *MOP:* equilateral triangle; *RAD:* 60° triangle
4. The isosceles, isosceles right, and equilateral triangles are on the top. The general, right, and 60° triangles are on the bottom, pretending to have two sides of the same length. See **Triangle Pretenders Present.gsp.**
5. The right, isosceles right, and 30°-60°-90° triangles are on the top. The general, isosceles, and 60° triangles are on the bottom, pretending to have right angles. See **Triangle Pretenders Present.gsp.**
6. Only the equilateral triangle is on the top. The general, isosceles, and 60° triangles are on the bottom, pretending to have three sides of equal length. See **Triangle Pretenders Present.gsp.**
7. Constructions will vary. The drag test will indicate whether students have successfully created a specific triangle or a pretender.

Triangle Pretenders

Name:

If you see a Sketchpad shape that looks like an isosceles triangle, how do you know whether it is constructed to always be isosceles? In this activity you'll use the drag test to find those shapes that are always what they appear to be and those that are only pretending.

EXPLORE

1. Open **Triangle Pretenders.gsp** and go to page "Equilateral Pretenders." Of the four shapes that look like equilateral triangles, only one is constructed to always have three sides of equal length. Drag the vertices to find it and match each shape with its most specific name.

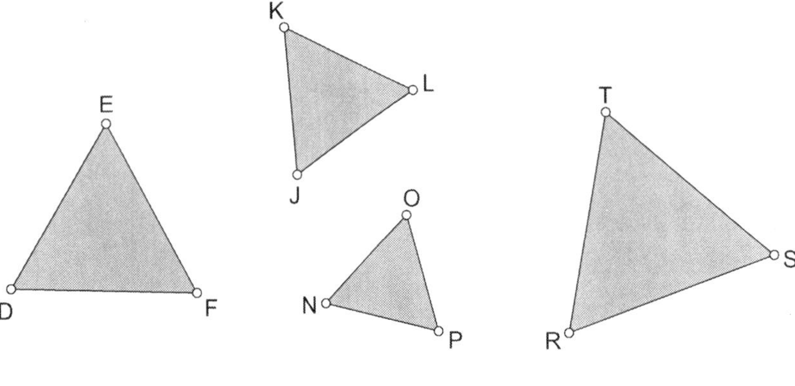

JKL	isosceles triangle
NOP	general triangle
DEF	equilateral triangle
RST	60° triangle

2. Go to page "Right Pretenders." Of the five shapes that look like right triangles, only two are constructed to always have a right angle. Find it and identify each shape with its most specific name.

ABC ____________________

EFG ____________________

MNO ____________________

RUS ____________________

HIT ____________________

Triangle Pretenders

continued

3. Go to page "Isosceles Pretenders." Of the six shapes that look like isosceles triangles, only some are constructed to always have two sides of equal length. Find them and match each shape with its most specific name.

BIG	general triangle
CUT	isosceles triangle
FLY	equilateral triangle
HEN	60° triangle
MOP	right triangle
RAD	isosceles right triangle

4. Go to page "Isosceles." Drag shapes that are always isosceles triangles to the top of the page. Drag shapes that can pretend to be isosceles triangles to the bottom of the page and make them look isosceles.

   To check that two sides have equal length, select them and choose **Measure | Length.**

5. Go to page "Right." Drag shapes that are always right triangles to the top of the page. Drag shapes that can pretend to be right triangles to the bottom of the page and make them look like right triangles.

   To check the right angles, select the three vertices with the right angle second and choose **Measure | Angle.**

6. Go to page "Equilateral." Drag shapes that are always equilateral triangles to the top of the page. Drag shapes that can pretend to be equilateral triangles to the bottom of the page and make them look equilateral.

EXPLORE MORE

7. Go to page "Make Your Own." Pick a type of triangle and construct it. Then make a pretender of that shape. For example, you might make an equilateral triangle and an isosceles triangle, or you might make a right triangle and a general triangle.

   Shape ________________

   Pretender's shape ________________

Three Pairs: Triangle Congruence Properties

ACTIVITY NOTES

INTRODUCE

Project the sketch for viewing by the class. Expect to spend about 10 minutes.

1. Open **Three Pairs.gsp** and go to page "Congruence." Enlarge the document window so it fills most of the screen.
2. Explain, ***Today you're going to use Sketchpad to explore what information is needed to prove that two triangles are congruent.*** If your students are not familiar with the word "prove," you might use the word "show" instead. ***What does it mean when I say two shapes are congruent?*** Students should reply that it means the figures are identical in size and shape. You might have volunteers follow the prompts on the sketch to demonstrate this. Make sure that point *A* is also dragged. Often students don't realize that congruent figures might be rotations or reflections, as well as translations of each other.
3. Write "$\triangle ABC \cong \triangle DEF$" on chart paper. ***What do you know about these two triangles from this statement?*** Here are some possible student observations.

   The ≅ sign means triangle ABC is congruent to triangle DEF. The two triangles are the same size and shape.

   The names of the triangles are written so that like parts are in the same order. Vertex A is like vertex D, vertex B is like vertex E, and vertex C is like vertex F.

   The related angles of the two triangles are congruent. Angle A is the same measure as angle D; angle B is the same measure as angle E; and angle C is the same measure as angle F.

   The related sides are congruent too. Side AB is congruent to side DE; side AC is congruent to side DF; and side BC is congruent to side EF.

Define *corresponding* as "matching" or "related," if needed.

4. ***When we talk about congruent triangles, there are six congruence statements we can make. The three sides of one triangle are congruent to the three sides of the other triangle. The three angles of one triangle are congruent to the three angles of the other triangle.*** Draw two congruent triangles, $\triangle ABC$ and $\triangle DEF$, on the chart paper. Use hatch marks and arcs to indicate the corresponding congruent parts. ***I can use hash marks to show the corresponding congruent sides and arcs with hash marks to indicate the corresponding congruent angles.*** Write the six congruence statements below your triangles: $\angle A \cong \angle D$; $\angle B \cong \angle E$; $\angle C \cong \angle F$; $\overline{AB} \cong \overline{DE}$; $\overline{AC} \cong \overline{DF}$; $\overline{BC} \cong \overline{EF}$.

5. ***Today you're going to explore whether you can use fewer than six congruence statements to prove that two triangles are congruent, and you're going to test which corresponding parts need to be congruent. First I'll show you where to find the sketch and how to use it.*** Model worksheet steps 3 and 4. Here are some tips.

   - Explain that each page of the sketch represents a different investigation.
   - In worksheet step 3, ask students to name the corresponding segments in the two figures. ***Which segment corresponds to segment AC? How do you know?*** [Segment *DF;* both segments have the same number of hash marks.] ***Which segment is congruent to segment EF?*** [Segment *BC*] Show how the corresponding side lengths remain equal when you change $\triangle ABC$.
   - In worksheet step 4, read the instructions on the sketch aloud. Model how to drag point *F* to construct $\triangle DEF$. Then demonstrate how to change the size and shape of $\triangle ABC$ by dragging its vertices. ***What happens when I change the size and shape of triangle ABC?*** Students should note that the corresponding congruent lengths in the second figure change so they remain congruent.

6. If you want students to save their work, demonstrate choosing **File | Save As,** and let them know how to name and where to save their files.

DEVELOP

Expect students at computers to spend about 20 minutes.

7. Assign students to computers and tell them where to locate **Three Pairs.gsp.** Distribute the worksheet. Tell students to work through step 12 and do the Explore More if they have time. Encourage students to ask their neighbors for help if they are having difficulty with Sketchpad.

8. Let pairs work at their own pace. As you circulate, here are some things to notice.

   - In worksheet steps 4–11, encourage students to change the shape and size of $\triangle ABC$ several times before making a conjecture for each congruence investigation. (Note that students will be "eyeballing" whether the two triangles are congruent. They won't be measuring lengths of sides and angles to prove whether the two triangles are congruent.)

Students can go back to the original setup for each investigation by choosing **Edit | Undo** several times, or they can hold down the Shift key and then choose **Edit | Undo All.**

- For each investigation, be sure students understand which corresponding parts are congruent in the two figures. In the second figure the congruent segments will be marked with the same number of hash marks and the congruent angles will be marked with the same colored arcs as in the first figure.

Order is important when naming the congruent parts and is implied by the order in which the letters are written.

- As students are writing their conjectures, watch for how they describe the sides and angles. Help students realize that their language needs to be clear. For the SAS investigation, for example, students who write, "If two sides and an angle of one triangle are congruent to two sides and an angle of another triangle, . . ." could be describing the SSA investigation. ***What angle are you referring to in your conjecture? How can you describe this angle so someone knows it's this one and not one of the others?*** If needed, introduce the terms *included angle* (an angle formed by two adjacent sides) and *included side* (a side that is common to two angles) to help students clarify their descriptions. These terms should be used relative to the including parts. ASA, for example, refers to two angles and *their* included side.
- Note that students tend to have a preconception that SSA guarantees congruence. That preconception may limit the exploration they do. Make sure that students realize that there are two possibilities, and take advantage of their surprise when they realize this.
- The investigation for AAS congruence has students explore only the case in which the corresponding sides are congruent. It is also possible for the noncorresponding sides to be congruent, which would not guarantee congruence. Emphasize the importance of using the term *corresponding* when writing the conjecture. This might also make a good extension. ***What if the congruent sides are not corresponding sides. Must the triangles still be congruent?***
- If students have time for the Explore More, they will investigate whether having two corresponding congruent parts is sufficient to guarantee that two triangles are congruent. Encourage students to make a prediction beforehand.

9. If students will save their work, remind them where to save it now.

SUMMARIZE

Project the sketch. Expect to spend about 15 minutes.

10. Gather the class. Students should have their worksheets with them. Ask, ***Are there shortcuts to determining triangle congruence?*** Then direct the discussion by creating on chart paper a table such as the one here, leaving the cells under the headings blank for the time being.

Congruent Parts	Triangles Always Congruent?	Congruence Statement
Side-Side-Side (SSS)	Yes	If three sides of one triangle are congruent to three sides of another triangle, then the triangles are congruent.
Side-Angle-Side (SAS)	Yes	If two sides and the included angle of one triangle are congruent to two sides and the included angle of another triangle, then the triangles are congruent.
Side-Side-Angle (SSA)	No	None
Angle-Side-Angle (ASA)	Yes	If two angles and the included side of one triangle are congruent to two angles and the included side of another triangle, then the triangles are congruent.
Angle-Angle-Side (AAS)	Yes	If two angles and a nonincluded side of one triangle are congruent to the corresponding angles and the corresponding nonincluded side of another triangle, then the triangles are congruent.
Angle-Angle-Angle (AAA)	No	None

11. Open **Three Pairs.gsp** and review each investigation. Have volunteers come up and explain their conjectures or show counterexamples using the sketch to reinforce their statements. Question students about their congruence statements if they are not detailed enough. Encourage students to use words such as *included* and *corresponding*. Fill in the chart with an agreed-upon congruence statement after each investigation is discussed.

12. When reviewing the AAA investigation, you may wish to ask students how the two triangles are related. The two triangles are similar, so they have the same shape, but not the same size.
13. Depending on the background of your students, you may wish to explain that the SSS, SAS, and ASA congruence statements are *postulates* (accepted as true without proof), whereas the AAS congruence statement is a *theorem* (proven to be true).
14. If time permits, discuss the Explore More. Students should find that two corresponding congruent parts in two triangles are not sufficient to guarantee that two triangles are congruent. ***Is knowing that two angles in one triangle are congruent to two angles in another triangle sufficient to state that the third angles are also congruent?*** [Yes; because the sum of the angle measures in a triangle is 180°, the third angle in both triangles is 180° minus the sum of the other two angles.] Students should recognize that this is like the AAA investigation, which they already determined could not guarantee triangle congruence.
15. You may wish to have students respond individually in writing to this prompt. ***Explain the difference between SAS and SSA. Can you use both to prove triangle congruence? Why or why not?***

EXTEND

Have students explore Hypotenuse-Leg (HL) congruence in right triangles. If the hypotenuse and a leg of a right triangle are congruent to the hypotenuse and a leg of another right triangle, are the triangles congruent? Let students try to construct two right triangles that are not congruent but have congruent hypotenuses and legs. Students should discover that only congruent triangles can be formed.

ANSWERS

2. All corresponding sides are congruent and all corresponding angles are congruent.
5. No. The three sides determine a unique triangle (or its reflection).
6. If two triangles have three pairs of congruent sides, then the triangles are congruent (SSS).

8. If two triangles have two pairs of congruent sides and a pair of congruent angles between the congruent sides, then the triangles are congruent (SAS).

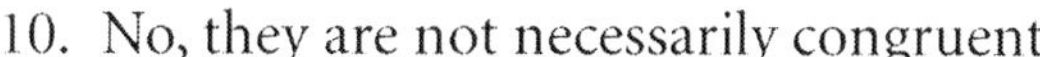

10. No, they are not necessarily congruent.

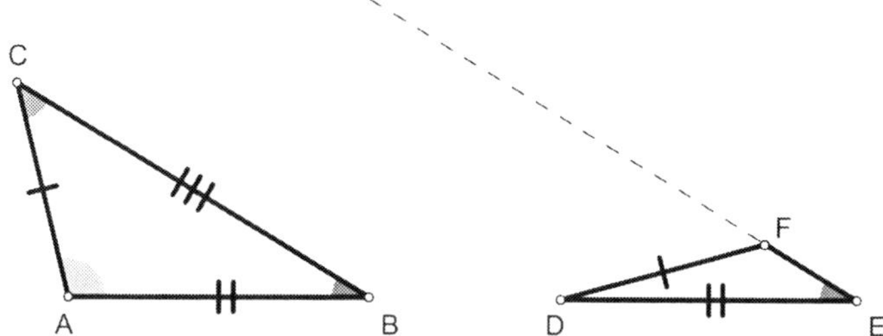

11. If two triangles have two pairs of congruent angles and a pair of congruent sides between the congruent angles, then the triangles are congruent (ASA).

    If two triangles have two pairs of congruent angles and a pair of congruent corresponding sides not between the congruent angles, then the triangles are congruent (AAS).

    AAA does not guarantee congruence.

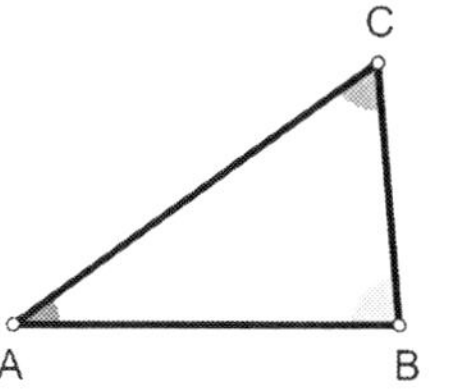

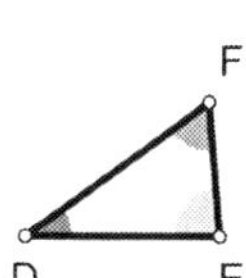

12. These combinations of parts guarantee congruence: SSS, SAS, ASA, and AAS. (For AAS, it's important to state that the sides must correspond. The correspondence is forced by the order in the other cases: It's impossible for the parts not to correspond.) SSA and AAA do not guarantee congruence.

13. Two corresponding congruent parts in two triangles are not sufficient to guarantee that the two triangles are congruent.

Three Pairs

Name:

Suppose the three sides of one triangle are congruent to three sides of another triangle. Are the two triangles congruent? What if two sides and the angle between them in one triangle are congruent to two sides and the angle between them in another triangle? Explore which combinations of sides and angles guarantee triangle congruence and which don't.

EXPLORE

1. Open **Three Pairs.gsp** and go to page "Congruence."

2. Read the text and follow the instructions. Explain what it means for two triangles to be congruent.

3. Go to page "SSS." You'll see a figure like the one shown and some text.

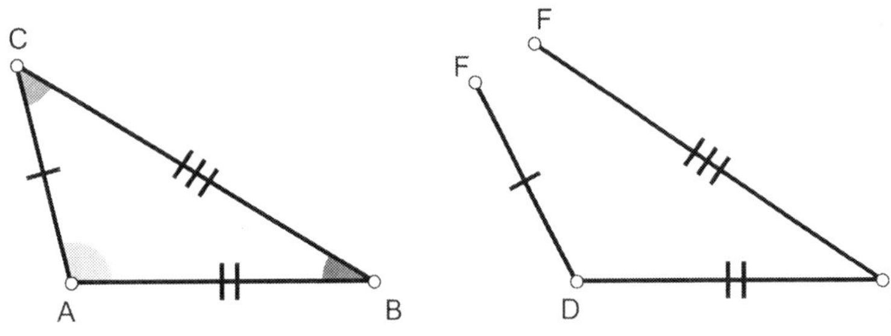

4. Read the text and follow the instructions to try to construct $\triangle DEF$ that is not congruent to $\triangle ABC$.

5. Could you construct a triangle with a different size or shape given three pairs of congruent sides? Explain.

6. If three sides of one triangle are congruent to three sides of another triangle (SSS), are the triangles necessarily congruent? If so, write a conjecture that summarizes your findings. If not, sketch a counterexample.

7. Go to page "SAS." Try to construct $\triangle DEF$ that is not congruent to $\triangle ABC$.

8. If two sides and the angle between them in one triangle are congruent to two sides and the angle between them in another triangle (SAS), are the triangles necessarily congruent? If so, write a conjecture that summarizes your findings. If not, sketch a counterexample.

9. Go to page "SSA." Try to construct $\triangle DEF$ that is not congruent to $\triangle ABC$.

10. If two sides and an angle not between them in one triangle are congruent to two sides and the corresponding angle not between them in another triangle (SSA), are the triangles necessarily congruent? If so, write a conjecture that summarizes your findings. If not, sketch a counterexample.

11. Go to each of the remaining three pages to explore the other triangle congruence investigations. For those that guarantee congruence, write a conjecture. For those that do not, sketch a counterexample.

12. Which combinations of corresponding parts (SSS, SAS, SSA, ASA, AAS, and AAA) guarantee congruence in a pair of triangles? Which do not?

EXPLORE MORE

13. Do you think that if two triangles have two corresponding parts congruent that you can guarantee the two triangles are congruent? Open a new sketch and try constructing two triangles that are *not* congruent for each set of congruent corresponding parts: Side-Side, Angle-Angle, and Side-Angle. Explain your findings.

Tiling with Triangles: Sums of Interior Angles in Polygons

INTRODUCE

Project the sketch for viewing by the class. Expect to spend about 10 minutes.

1. Open **Tiling with Triangles.gsp.** Go to page "Triangle." Drag a vertex or side of the triangle to change its shape, and ask students to watch the measurements update. Press *Sum of Measures of Two Angles* to show that the sum of two angles (B and C) updates continuously as a triangle vertex is dragged. You might model measuring two other angles and show that their sum updates, and do the same for the third pair. Then use the other button to show the angle measurement for the sum of three angles. As the shape of the triangle changes, it looks like that sum is staying the same. Ask, ***Does it update as the triangle is changed?*** Elicit from students that the sum *is* updating: It appears to stay the same because the sum of the measures of the interior angles of any triangle is always the same—180°.

2. Go to page "Quadrilateral." ***What might the sum of the interior angles of a quadrilateral be? How could you find the sum of the angles without measuring them?*** Have students discuss their ideas with someone sitting next to them. If students know already that the sum of the measures of the angles for a quadrilateral is 360°, ask them to explain how they know.

3. Solicit responses. If students don't suggest showing two triangles within the quadrilateral, ask, ***How could you use what you know about triangles?***

   Allow time for students to suggest drawing a diagonal to form two triangles. Model using the **Segment** tool to draw a line segment connecting one pair of opposite vertices in the quadrilateral. Facilitate discussion so that all students recognize that the sum of the angles must be 360°, twice the sum of the angles of a triangle.

For students new to Sketchpad, it is essential that they understand that the points at which the segment begins and ends must highlight as the **Segment** tool attaches the segment to the point.

4. Start a table on the board. You might ask students to add to it or make their own tables as they explore polygons with more sides.

5. ***You're going to explore the sum of the measures of the interior angles for other polygons.*** Clarify that students should work with convex polygons only. Drag a vertex of the quadrilateral to make a concave figure, then return it to a convex quadrilateral.

DEVELOP

Expect students at computers to spend 25 minutes.

6. Assign students to computers and distribute the worksheet. Explain that students should work the problems through worksheet step 8. If they finish the 8 steps, they can do the Explore More. Steps 9–12 can be completed during Summarize.

7. As students work, the question of how to draw the triangles may come up. You may want to hold a check-in with the class. Elicit a clear description, such as the following student example.

   Pick one vertex. Connect it to all the other vertices. Two vertices are already connected to the starting vertex as the sides of the polygon. When you draw all the triangle sides from one vertex, it will keep you from drawing a line segment that crosses another segment.

Clarify that as students drag a polygon, they should avoid causing any two sides to form a line or overlap other sides.

Refrain from suggesting that students record their findings in a table at this point. Listen, instead, for how students' thinking develops as they explore the sequence of polygons.

8. As you circulate, make sure students dragging polygons avoid making any two sides form a line or overlap other sides. Listen for students' observations, conjectures, and predictions about possible patterns and relationships. Here are samples of students' thinking.

   How many triangles are there? Is it two fewer than the number of sides again?

   Every time there's one more side, there's another 180 degrees.

   Ask students to add to the table and let selected students know that you will want them to share their thinking.

9. In worksheet step 8, the question will prompt students to consider relationships among the number of sides, number of triangles, and sum of interior angle measures. Some students may already have ideas about or feel confident in these relationships. For others, the questions will focus their thinking and prepare them for the class's work together in the summarizing discussion on steps 9–12.

SUMMARIZE

Working away from computers, expect to spend about 25 minutes.

10. Bring the class together. Students should have their worksheets with them. ***Talk with a partner about what you were doing to find the sum of the angle measures for the polygons. Were you doing something similar each time?***

11. Ask students to share their thinking. Students may suggest these ways.

    We multiplied the number of triangles by 180.

    We subtracted 2 from the number of sides of the polygon and multiplied that number by 180.

    Students may question whether the two approaches are really different. In response, some students may make arguments such as the ones that follow. It's not necessary that students resolve this question at this time. Ask them to keep it in mind as they continue working.

    You can't know the number of triangles without knowing the number of sides.

    If you know the number of sides, that's all you need to know.

    You need not resolve this point. Instead, help students consider how the angle sum can always be found.

Creating a Verbal Representation

12. ***How would you tell someone else how to find the sum of the interior angle measures for any polygon without drawing triangles?*** Record the students' expressions as they propose and refine a description such as this sample.

    Start with the number of sides of the polygon. Subtract 2 (to get the number of triangles). Multiply the result by 180 (for the number of degrees in one triangle).

13. Work with a specific example. ***Let's take the example of a polygon with 12 sides. How would you find the total angle sum using the description you just wrote?*** As students respond that they would start with 12 for the number of sides and subtract 2, record the expression.

    $$12 - 2$$

    As students continue that they would multiply the result by 180, finish recording the expression.

    $$(12 - 2)180$$

Asking whether a rule or strategy "works" for more cases, and for all cases, is an important element of learning to generalize, a basic part of algebraic thinking.

14. ***Can we use this way of recording to find the sum of the angle measures for more polygons—without having to draw triangles?*** Note students' first responses: Are some nodding and others unsure? ***What about a polygon with 22 sides? Think about this with your neighbor and be***

ready to discuss. When students have an answer, ask, ***What did you find out?*** Let students come to agreement that the recording (22 − 2)180 "works." (Students should be able to arrive at the sum, 3600°, using mental computation.)

15. ***How would you find the sum for a polygon with 50 sides? For one with 100 sides? What recordings would you make?*** Record students' responses as they supply the expressions.

    (50 − 2)180 (100 − 2)180

Developing an Algebraic Expression

16. Call students' attention to the expressions you recorded for the polygons with 12, 50, and 100 sides. ***What stays the same in each of these expressions?*** Students should recognize that "the 2" and "the 180" stay the same. ***What does each of those numbers tell about how to find the sum of the measures?*** Students may share ideas such as these.

    It's always true that you subtract 2 from the number of sides because there are two fewer triangles than sides.

    The number of degrees in a triangle is always 180, and you always multiply by the number of degrees in a triangle.

17. ***Suppose we wanted to write this expression so it showed how to find the sum of the interior angle measures for any polygon. Mathematicians have a way of doing this. How might you do it?*** Have students work alone or in pairs to try to write an expression to answer question 12. Students are likely to think of using letters to stand for numbers. Take suggestions.

18. Invite students to share their ideas, and guide the class in incorporating a variable in the expression they have been using. The following ideas about the expressions should arise in the discussion.

    - The number that does not apply to all polygons is the one for the number of sides. That number can change because it depends on how many sides a polygon has.
    - Two of the numbers in the expression apply to all polygons. The relationships they represent don't change.
    - A letter can be used to represent the number that changes. The term for a letter that represents a number that can change is *variable.*

- The number of sides is the variable in this expression.
- Students might have chosen n (for number of sides), s (for sides), x (because they have seen it used before), or another letter.
- We can use a variable for the number of sides in the expressions. The result is an algebraic expression in the form $(n - 2)180$.

Here the variable varies. For any whole number n, the expression $(n - 2)180$ gives the number of degrees in the sum of the measures of angles of a polygon with n sides.

Some students will bring the misconception that a variable must be the unknown in an equation—that is, a variable represents the one number that solves an equation. ($x + 5 = 7$, so $x = 2$; x does not vary.) Here *variable* is used as a quantity that can vary.

19. Now have the class use the expression (with any letter for the variable) to solve problems you or volunteers pose. ***A polygon has 17 sides. What is the sum of its interior angle measures?*** [(17 − 2)180°, or 2700°]

Explore More

20. Step 13 on the worksheet presents a problem in a form that requires students to work backward: They are given the sum of the measures of the interior angles for a polygon and are asked to determine the number of sides the polygon has. Presenting the same situation from different angles is a good way to build students' ability to use algebraic thinking.

EXTEND

Students who have extra time might enjoy adding to the sketches in **Tiling with Triangles.gsp** the sum of the measures of the angles for the various polygons. Sketchpad can measure angles and calculate their sum.

Model these steps.

Click on a clear space before you select the points to see the choice **Measure | Angle.**

- For each angle, select three points, with the vertex your middle selection, and choose **Measure | Angle.**

If the Calculator is in the way, move it by dragging the title bar.

- Choose **Number | Calculate** to display the Calculator. Click once on each measurement to enter it into a calculation, entering plus signs between the measurements.

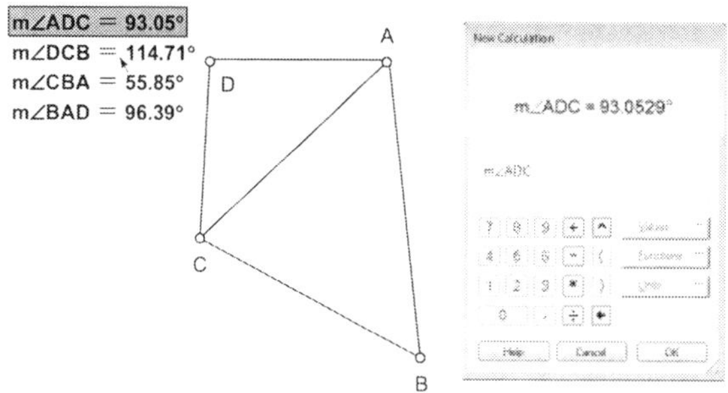

- Click **OK.**

ANSWERS

3. a. 3
   b. No
   c. 540°
4. Predictions may vary.
5. a. 4
   b. 720°
6. Predictions may vary.
7. Predictions may vary.
8. Answers will vary, depending on students' predictions.
9. Disagree. The number of degrees is equal to the number of triangles times 180. The number of triangles is 2 less than the number of sides, so 180° is multiplied by the number of sides minus 2.
10. Sample answer: I'd subtract 2 from 22 to find the number of triangles and then multiply 20 times 180.
11. $(22 - 2)180$
12. $(n - 2)180$
13. The polygon has 12 sides. Sample reasoning: The total, 1800°, is how many times 180? Ten times. So, the number of triangles is 10. The polygon has 2 more sides than the number of triangles, so I know the polygon has 12 sides.

Tiling with Triangles

Name:

In this activity you will explore the sum of the measures of the interior angles of polygons.

1. Open **Tiling with Triangles.gsp.** Go to page "Quadrilateral."

2. Create two triangles whose angles equal the quadrilateral's angles.

3. Go to page "Pentagon."

   Draw line segments to connect vertices. Make sure the segments do not intersect.

   a. How many triangles can you make? ______________________

   b. Drag vertices. Does the number of triangles stay the same? __________

   Compare the angles of the triangles to the angles of the pentagon.

   c. What is the sum of the measures of a pentagon's interior angles? ________

4. Predict: How many triangles can be shown in a hexagon? ______________

5. Go to page "Hexagon." Draw nonintersecting segments to connect vertices.

   a. How many triangles can you make? ______________________

   b. What is the sum of the interior angle measures? ______________________

6. Think about a heptagon and make predictions.

   a. How many triangles will you be able to make in the heptagon? __________

   b. What will be the sum of the interior angle measures? ______________

7. Make predictions for an octagon.

   a. Number of triangles ______________________________

   b. Sum of the interior angle measures ______________________

8. Go to page "Heptagon" and then to page "Octagon." Form triangles to test your predictions. Are your predictions correct?

Tiling with Triangles

continued

9. Carmen said the number of degrees is equal to the number of sides multiplied by 180. Do you agree? Why? If you don't agree, what would you say? Explain your thinking.

10. What would you say to describe how you would find the sum of the measure of the angles for a polygon with 22 sides?

11. Write your description as a number expression. ______________

12. Mathematicians write an algebraic expression to describe the sum of the interior angles of any polygon. Talk with your teacher and others. Develop a mathematical expression for the sum of the measures of the interior angles of a polygon with *n* sides.

EXPLORE MORE

13. Kate says, "I'm thinking of a polygon. The sum of the measures of its interior angles is 1800°. How many sides does the polygon have?"

    Solve Kate's problem any way you want. If you want to draw polygons, use page "Explore More" of the sketch. Record your work and answer here.

Small, Medium, and Large: Triangle Inequalities

ACTIVITY NOTES

INTRODUCE

Project the sketch for viewing by the class. Expect to spend about 10 minutes.

1. Open Sketchpad and enlarge the document window so it fills most of the screen. Explain, ***Today you are going to use Sketchpad to discover some relationships among the measures of the sides and angles in a triangle. I will demonstrate how to construct a triangle, measure a side length, and find the sum of two side lengths. Then I will show you how to measure an angle. After I'm done, you will construct a triangle and measure all side lengths and all angles, and explore their relationships. Based on your observations, you will make some conjectures.***

2. As you demonstrate, make lines thick and labels large for visibility. First, model the triangle construction in worksheet step 1. Then model how to measure two side lengths and how to find the sum of their measurements in worksheet steps 2 and 3. Finally, model how to measure an angle in step 8. Here are some tips.

   - In worksheet step 1, use the **Segment** tool to construct the three sides of the triangle. After constructing the first side, click on one of its endpoints to construct the second side. Do the same to attach the third side to the second side. Complete the construction by clicking on the second endpoint of the second side and the first endpoint of the first side.

   - Use the **Arrow** tool to drag the vertices to be sure the sides are connected. ***Test your construction by dragging the vertices.*** Stress the importance of the drag test. Model how to choose **Edit | Undo** to show students what to do when they make an error in their constructions.

   - Model how to choose the **Text** tool to label the three vertices by clicking on each point. Explain that students can double-click a label to rename it if needed.

You can set the measurement units and precision by choosing **Edit | Preferences | Units.**

   - In worksheet step 2, model how to measure the length of a side. Select a side (a segment), and choose **Measure | Length.** Explain that clicking in any blank space after each measurement will deselect objects before they make the next measurement. Read the measurement that appears in the document window. ***The* m *in front of the segment stands for "the measure of."***

If students make errors in their calculations, they can double-click the measurement to edit it.

- Measure the length of another side, and then model worksheet step 3 to find the sum of the two side lengths. Choose **Number | Calculate,** and then click on a measurement to make it appear in the Calculator. Use the Calculator keypad for the + sign. After you click **OK,** the measurement will appear on the sketch. Read the measurement aloud to the class using the phrase *the measurement of* when you come to *m* before the endpoints that name a segment.
- For worksheet step 8, model how to measure one angle by selecting the three points that define it. ***When measuring an angle, always select the vertex as the second point.*** Identify the vertex as the point where the two sides of the angle meet. Choose **Measure | Angle** and read the measurement aloud.

3. If you want students to save their work, demonstrate choosing **File | Save As,** and let them know how to name and where to save their files.

DEVELOP

Expect students at computers to spend about 20 minutes.

4. Assign students to computers and distribute the worksheet. Tell students to work through worksheet step 11 and do the Explore More if they have time. Encourage students to ask their neighbors for help if they are having difficulty with the construction.
5. Let pairs work at their own pace. As you circulate, here are some things to notice.
   - In worksheet step 4, as students drag the vertex, ask them what they notice. Students may reply using language such as, *The measurements aren't close,* or *The measurements are not the same.* Ask students to make more descriptive statements. ***How is the sum of the two side lengths related to the third side? How does this change as you drag a vertex?***
   - In worksheet step 5, when the sum of the lengths of two sides equals the length of the third side, the figure becomes a segment. Watch for students who say that it is possible for the sum of the two side lengths in a triangle to be equal to the third side. ***What happened when the two measurements were equal? How would you describe this figure?***

- In worksheet step 6, students won't be able to drag the vertex to make the sum of the two side lengths less than the third side. Encourage students to test whether this holds true for any two side lengths and the related third side length.
- In worksheet step 7, ask students to state the conjecture two ways. ***Can you state your conjecture another way?*** [The length of any one side of a triangle must be less than the sum of the lengths of the other two sides. The sum of the lengths of two sides of a triangle must be greater than the length of the third side.]
- For worksheet steps 9–11, some students may not understand which angle is opposite which side. ***Think about extending an arrow from the vertex of an angle through the center of the triangle. If the arrow continued, which side would it hit?*** Ask students to share some of their observations. Listen for these ideas.

  If I make a side length smaller, the angle opposite becomes smaller.

  If I make an angle larger, the side opposite becomes larger.

  The shortest side and the middle length side form the sides of the largest angle.

  I can't make a triangle that has the longest side opposite the smallest angle.

  If I make a triangle with two sides of equal length, the angles opposite have equal measure.

  If I make a triangle with all three sides the same length, then all three angles are the same measure.
- If students have difficulty seeing the relationships, ask them to describe the location of the longest side and the largest angle.

6. If students have time for the Explore More, they will construct the medians of a triangle, measure their lengths, and then observe the relationship between the lengths of the sides and the lengths of the medians.
7. If students will save their work, remind them where to save it now.

SUMMARIZE

Project the sketch. Expect to spend about 15 minutes.

8. Gather the class. Students should have their worksheets with them. Open **Small Medium Large Present.gsp** and use the pages as needed. Begin the discussion by asking students to review worksheet step 5. Have a volunteer come up and demonstrate what happens when the lengths of two sides of the triangle equal the length of the third side. [The triangle collapses and becomes a segment.]

9. For worksheet step 6, ask students to say what they discovered. ***Was it possible for the sum of the two side lengths to be less than the third side length?*** [No] ***What do you think happens when you try to make a triangle?*** [Two of the sides don't reach.]

10. For worksheet step 7, write students' conjectures on chart paper. Help students come to an agreement about common wording. ***Can you use symbols to state this?*** [For any triangle ABC, $mAB + mBC > mCA$, $mAB + mCA > mBC$, and $mBC + mCA > mAB$.] ***This is known as the*** **Triangle Inequality Theorem.**

11. You may wish to have students respond individually in writing to this prompt. ***Based on your conjecture, do you think you can use any three side lengths to form a triangle? Explain and use an example to support your explanation.*** [Examples will vary. No, if you have side lengths of 2, 3, and 6, they won't make a triangle because $2 + 3 < 6$.]

12. In worksheet step 11, have students state their conjectures and write them on chart paper. Work with the class to create the best wording. Ask whether students can state the conjecture symbolically. [For triangle ABC with $m\angle A > m\angle B > m\angle C$, then $m\overline{BC} > m\overline{AC} > m\overline{AB}$.]

13. You may wish to have students respond individually in writing to this prompt. ***Based on your conjecture, do you think a triangle with three different side lengths can have two angles with the same measure? Explain your reasoning.*** [No, if you have three sides of different lengths, the angles opposite will have different measures. The longest side will be opposite the greatest angle measure; the shortest side will be opposite the least angle measure; and the middle side length will be opposite the middle angle measure.]

14. If time permits, discuss the Explore More. Have students share their findings.

ANSWERS

7. Sample answer: If the sum of two side lengths equals the third side length, two sides of the triangle collapse onto the third side, forming a segment instead of a triangle.

8. Sample answer: It's impossible for the sum of the lengths of any two sides of a triangle to be less than the length of the third side. If the sum of the lengths of the two segments (sides) is less than the length of the third side, the two short segments together are not long enough to meet and create a closed shape.

9. Sample answer: The length of any one side of a triangle must be less than the sum of the lengths of the other two sides.

12. Students should fill in the chart with these angles.

Longest Side	Largest Angle	Shortest Side	Smallest Angle
$\overline{AB}$	$\angle C$	$\overline{AB}$	$\angle C$
$\overline{AC}$	$\angle B$	$\overline{AC}$	$\angle B$
$\overline{BC}$	$\angle A$	$\overline{BC}$	$\angle A$

13. Sample answer: The largest angle in a triangle is always opposite the longest side, and the smallest angle is always opposite the shortest side.

14. Sample answer: The longest median is to the shortest side. The shortest median is to the longest side.

Small, Medium, and Large

Name:

In this activity you'll explore the relationships among the measures of the sides and angles in a triangle.

CONSTRUCT

1. In a new sketch, construct a triangle.

2. Label the vertices by clicking them in order from *A* through *C*.

3. Drag each of the three vertices to make sure the three sides are connected.

4. Measure the length of the three sides. To measure the length of a segment, select it and choose **Measure | Length.** Click in any blank space to deselect all objects after each measurement.

5. Now you'll calculate the sum of any two side lengths.

   Choose **Number | Calculate** to show Sketchpad's Calculator.

   Click once on a measurement to insert it into a calculation.

   Use the + on the Calculator keypad to find sums.

   When you are done, click **OK.**

m $\overline{AB}$ = 4.81 cm
m $\overline{BC}$ = 3.54 cm
m $\overline{CA}$ = 5.15 cm

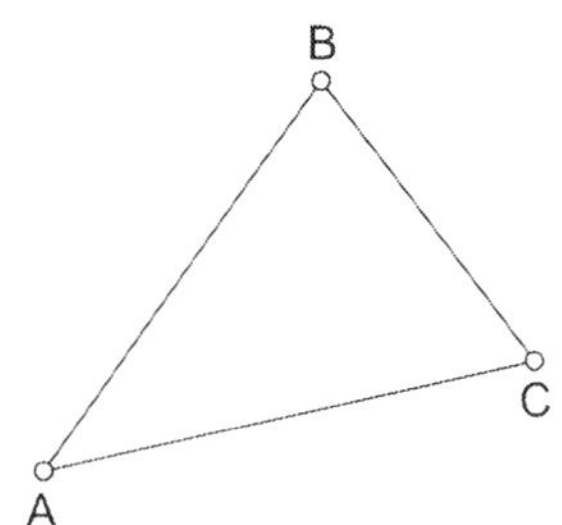

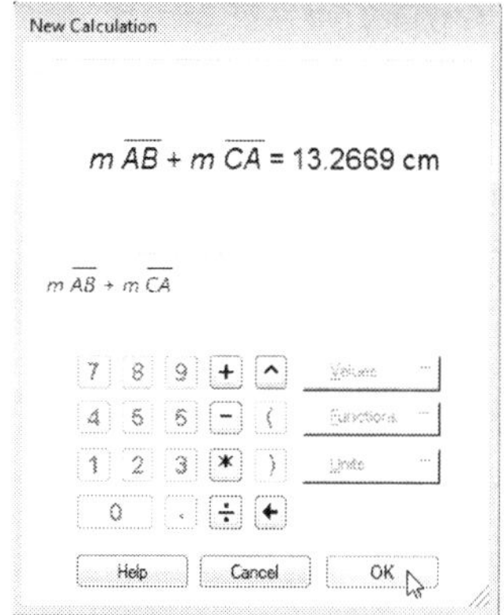

EXPLORE

6. Drag a vertex of the triangle to try to make the sum you calculated equal to the length of the third side.

Small, Medium, and Large

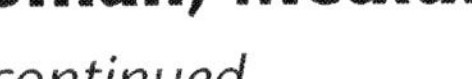

continued

7. Is it possible for the sum of two side lengths in a triangle to be equal to the third side length? Explain.

8. Do you think it's possible for the sum of the lengths of any two sides of a triangle to be less than the length of the third side? Explain.

9. Summarize your findings as a conjecture about the sum of the lengths of any two sides of a triangle.

10. Measure $\angle ABC$, $\angle BAC$, and $\angle ACB$. To measure an angle, select three points, with the vertex as your second point. Then choose **Measure | Angle.** Click in any blank space to deselect all objects after each measure.

11. Drag the vertices of your triangle and look for relationships between side lengths and angle measures.

12. In each area of the chart below, a longest or shortest side is given. Fill in the chart with the name of the angle with the greatest or smallest measure, given that longest or shortest side.

Longest Side	**Largest Angle**	**Shortest Side**	**Smallest Angle**
$\overline{AB}$		$\overline{AB}$	
$\overline{AC}$		$\overline{AC}$	
$\overline{BC}$		$\overline{BC}$	

13. Summarize your findings from the chart as a conjecture.

EXPLORE MORE

14. A *median* of a triangle is a line segment drawn from one vertex to the midpoint of the opposite side. Use your triangle from step 1 to explore the relationship between the side lengths of a triangle and the lengths of its medians.

    To find the midpoint of a side, select the side and choose **Construct | Midpoint.**

    To construct a median, select the midpoint and the vertex opposite it and choose **Construct | Segment.**

    To measure a median, select the median and choose **Measure | Length.**

    When you are done, drag the vertices. What do you notice about the lengths of the sides and the lengths of the medians?

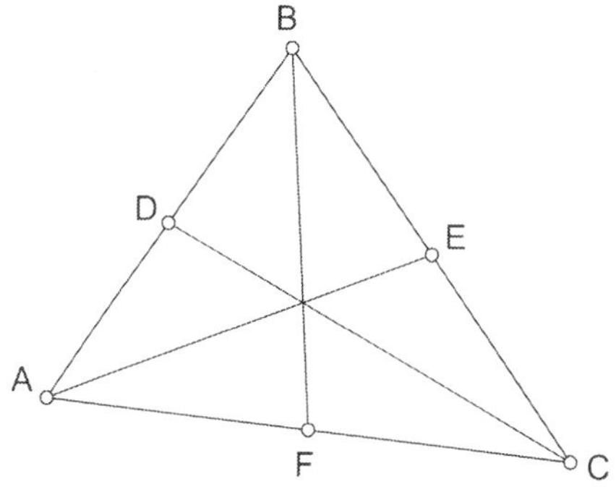

INTRODUCE

Project the sketch for viewing by the class. Expect to spend about 5 minutes.

1. Open **Tilted Squares Present.gsp** and go to page "Centimeter Grid." Enlarge the document window so it fills most of the screen. Explain, ***Today you are going to use Sketchpad to find square roots of numbers. You will construct squares and measure their areas and side lengths. Many of your measurements will be approximations. Sketchpad will round your measurements to the nearest thousandth of a centimeter.***
2. As you demonstrate, make lines thick and labels large for visibility. Model worksheet steps 1–4. Here are some tips.
   - ***You will use the Sketchpad centimeter grid like a geoboard or a piece of dot paper.*** Use the **Segment** tool to construct a segment to demonstrate how the endpoints snap to a point on the grid.
   - In step 2, explain that students will be using a custom tool to construct squares. ***Instead of constructing a square, you will use a custom tool that constructs squares. This tool will save you time, so you will be able to measure the area and side length of many different-sized squares quickly.*** Demonstrate how to choose **Square** from the Custom Tools menu and model how to construct a square.
3. If you want students to save their work, demonstrate choosing **File | Save As,** and let them know how to name and where to save their files.

DEVELOP

Expect students at computers to spend about 15 minutes.

4. Assign students to computers and tell them where to locate **Tilted Squares.gsp.** Distribute the worksheet. Tell students to work through step 13 and do the Explore More if they have time. Encourage students to ask their neighbors for help if they are having difficulty using Sketchpad for the construction.
5. Let pairs work at their own pace. As you circulate, here are some things to notice.
   - In worksheet step 7, listen for students' responses. Students may focus on the fact that the side length multiplied by the side length, or the side length squared, equals the area. They may not note that the inverse operation is true as well: The side length is the square root of the area. ***If you are given the area of a square, how can you find its side length?***

- In worksheet step 11, if students drag some of the vertices, the square will slide (translate) along the grid, not change size. ***Try dragging a different vertex to change the size of the square.*** If students mistakenly make the same entry twice, they can remove the most recently added row from a table by pressing the Shift key while double-clicking the table.
- Look for students who are only constructing squares with side lengths along the horizontal and vertical grid points. ***Can you construct a square that is tilted a different way?***
- In worksheet step 12, watch for students who don't understand what to do. ***How can you find a square root using the area of the square?*** [The square root is the length of the side of the square.] ***How many squares will you need to construct to find 12 different square roots?*** [At least 12 different-sized squares] ***Can you add a square with an area of 25 cm^2 to your table?*** [No, the areas need to be less than or equal to 20 cm^2.]
- In worksheet step 13, observe students who quickly answer "yes." ***What do you notice about all the area measurements in your table?*** [They are all a whole number of square centimeters.] ***Can you construct squares with areas of whole number square centimeters from 1 cm^2 to 5 cm^2?*** Let students explore this for a while. They will not be able to construct a square with an area of 3 cm^2 on a square grid.
- If students have time for the Explore More, they will plot a table of measurements on a coordinate grid and use their graphs to estimate other square roots. ***What do the x- and y-coordinates represent?*** [The x-coordinates represent the squares and the y-coordinates represent the square roots.]

6. If students will save their work, remind them where to save it now.

SUMMARIZE

Project the sketch. Expect to spend about 10 minutes.

7. Gather the class. Students should have their worksheets with them. Open **Tilted Squares Present.gsp** and use it to support the class discussion. Drag a vertex to construct a square with side length 3 cm.

8. Start off by explaining that any number raised to a power of 2, or "squared," can be represented as a geometric figure—a square. ***What number squared is represented by the square figure on this sketch?*** [3] ***What does the 3 represent?*** [The length of a side of the square, 3 cm] ***What is 3^2?*** [9] ***What does the 9 represent?*** [The area of the square, 9 cm^2] ***What is the formula for the area of a square?*** [$s \times s$, or s^2] ***How can you use the area of this square to find the length of a side?*** [Find the square root of the area.]

9. Introduce the square root symbol and read $\sqrt{9} = 3$ as "the square root of 9 equals 3." ***The square root of a number is the inverse operation to the square of a number.***

10. ***Let's see how many square roots we can find for numbers 20 or less.*** Let volunteers come up, construct a square, and add its measurements to the table.

11. Discuss the table when the class is done constructing squares. ***What do you notice about the area measurements?*** [They are all whole number measurements.] ***What do you notice about the side length measurements?*** [Some are whole number measurements and some are not.] Explain that when you find the square root of a number and it is a whole number, the original number is called a *perfect square*. ***Which numbers are perfect squares?*** [1, 4, 9, 16] ***What would be the next perfect square? How do you know?*** [It would be the next whole number squared, $5^2 = 25$.]

12. ***According to the table, what is the square root of 2?*** [1.414] ***If you squared 1.414, do you think it would equal 2?*** Explain. [No, 1.414 is an approximation to the nearest thousandth.]

13. In worksheet step 13, listen to students' explanations. ***Were you able to find the square root of any number using this method? Explain.*** [No, it's not possible to find the square roots of 3, 6, 7, 11, 12, 14, 15, or 19 using a square grid.]

14. If time permits, discuss the Explore More. ***How can you use the graph to estimate square roots?*** [Find the value of the square on the x-axis. Find the corresponding y-coordinate of the point on the line through the plotted points from the table. The y-coordinate is the approximate square root.]

EXTEND

What questions occurred to you during this exploration? Encourage curiosity. Here are some sample student queries.

Is the side length of a cube the cube root of its volume?

Are all the non-whole number square roots irrational numbers?

Is there a pattern in the order of the perfect squares?

ANSWERS

7. A square's side length is the square root of its area.

12. Students should be able to find exactly 12 square roots of whole numbers less than 20. If you include the square with a side length of 0, there are actually 13 possible table entries. Their tables should contain these squares and corresponding square roots.

Square	0	1	2	4	5	8	9	10	13	16	17	18	20
Square root	0	1	1.414	2	2.236	2.828	3	3.162	3.606	4	4.123	4.243	4.472

13. It is impossible to find the square root of every whole number using squares drawn on a square grid. For example, it is impossible to draw a square with area 7 square units on a grid.

14. The plot should look like this.

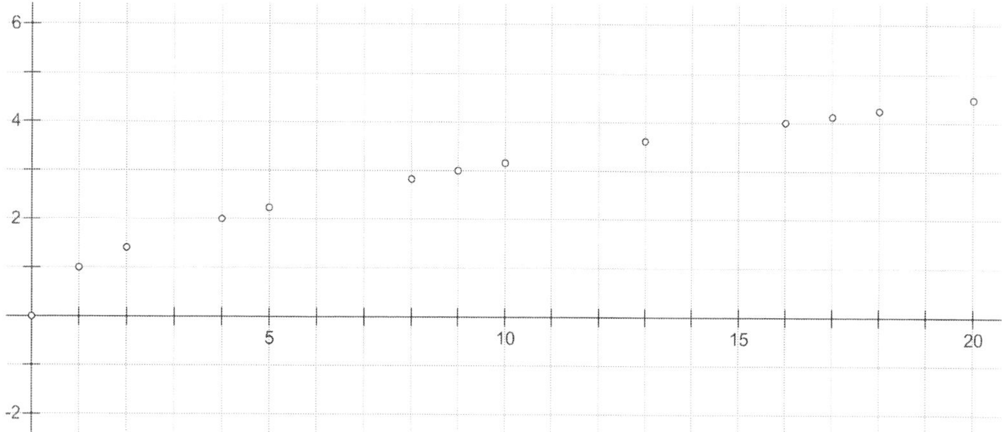

Student estimates for $\sqrt{3}$ should be close to 1.7, for $\sqrt{7}$ should be close to 2.6, and for $\sqrt{12}$ should be close to 3.5.

Tilted Squares

Name:

In this activity you'll construct squares on a grid to find the square roots of numbers. If you know the area of a square, you can find the square root of its area by measuring one of its sides.

CONSTRUCT

1. Open **Tilted Squares.gsp** and go to page "Centimeter Grid." Any points you construct on this page will snap to the grid.

2. You'll use a custom tool to construct squares. Press and hold the **Custom** tool icon and choose **Square** from the Custom Tools menu.

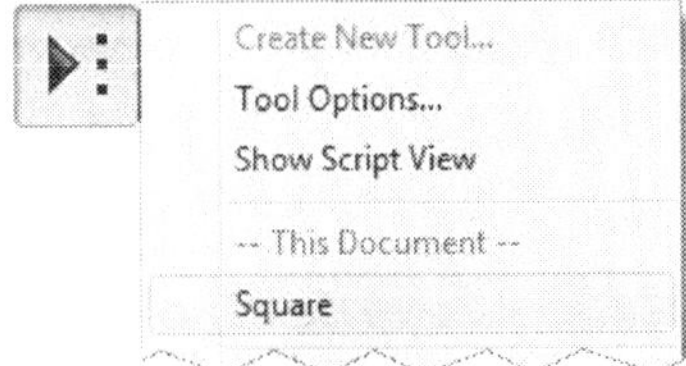

3. Use the custom tool to construct a square.

4. Click in blank space to deselect all objects. Drag the vertices of the square in all direction to observe how it behaves. Make sure to tilt the square.

EXPLORE

5. Select the interior of the square and choose **Measure | Area.**

6. Select one side of the square and choose **Measure | Length.**

7. How is a square's side length related to its area?

8. Select in order the area measurement and the length measurement and choose **Number | Tabulate.**

9. Drag a vertex to make the square a different size.

10. Add another entry to your table by double-clicking the table.

11. Continue adding entries to your square-root table by changing the size of your square and double-clicking inside the table.

12. Find the square roots of 12 different numbers less than or equal to 20. Record the numbers and their square roots here.

Square						
Square root						

Square						
Square root						

13. Do you think it is possible to find the square root of any whole number using this method? Explain your reasoning.

EXPLORE MORE

14. Now you'll plot your table of measurements on a coordinate grid and use your graph to estimate other square roots.

Click on any dot to select the grid.

Go to the Graph menu, and choose **Dotted Grid** to uncheck it. (Grid lines will appear in place of the grid dots.) Then press the *Show Axes* button.

Select the table and choose **Graph | Plot Table Data.** Click **OK.**

Use your graph to estimate $\sqrt{3}$, $\sqrt{7}$, and $\sqrt{12}$.

Squaring the Sides: The Pythagorean Theorem

INTRODUCE

Project the sketch for viewing by the class. Expect to spend about 10 minutes.

1. Open Sketchpad and enlarge the document window so it fills most of the screen. If you are short on time, you can have students use **Squaring the Sides.gsp,** which already contains the **Square** custom tool that students create in worksheet steps 2–9.

2. Explain, ***Today you're going to use Sketchpad to explore the Pythagorean Theorem. You will create a custom tool that constructs squares and use it to construct squares on the sides of a right triangle. Then you'll measure lengths of the sides of the right triangle and the areas of the attached squares and make a conjecture about their relationship. Before you start, I'll model how to make the custom tool.*** As you demonstrate, make lines thick and labels large for visibility. Model how to create the **Square** custom tool. Here are some tips.

   - In worksheet step 1, model how to set the Sketchpad Preferences so that points are automatically labeled. ***For this sketch, we'll have Sketchpad automatically label the points as we construct them. It'll be easier to follow the construction.***

   - In worksheet step 3, explain to students that the marked point will flash briefly to indicate it has been marked. ***Point A will be used as a center for the rotation.*** Tell students that in the rotated image, the point corresponding to point B is B' and is read "B prime." ***What type of angle is $\angle B'AB$? How do you know?*** [It is a right angle because $\overline{AB}$ was rotated 90°.]

   - In worksheet step 4, model the rotation and again ask students to name the type of angle formed. ***What type of angle is $\angle AB'A'$?*** [Right angle]

   - In worksheet step 5, it may be easier to construct the segment by selecting point A' and point B and choosing **Construct | Segment.**

   - In worksheet step 6, remind students that squares are named by listing their vertices in consecutive order. ***What is this shape called?*** [Square $ABA'B'$ or square $BAB'A'$]

   - In worksheet step 7, model how to drag each vertex to test the construction. ***It is important not to skip this step. You need to test your construction to make sure it holds together.*** Model how to choose **Edit | Undo** if students make mistakes.

- In worksheet step 9, model how to press and hold the **Custom** tool icon to display the Custom Tools menu. When you are done naming the custom tool, press and hold the **Custom** tool icon to show students how the **Square** tool now appears in the Custom Tools menu. ***You can now use the Square custom tool to make other squares.*** Show students how to select and use the **Square** tool.
- You might also model how to add a new blank page to a sketch, as described in worksheet step 13.

3. If you want students to save their work, demonstrate choosing **File | Save As,** and let them know how to name and where to save their files.

DEVELOP

Expect students at computers to spend about 30 minutes.

4. Assign students to computers. Distribute the worksheet. Tell students to work through step 29 and do the Explore More if they have time. Encourage students to ask their neighbors for help if they are having difficulty with the construction.
5. Let pairs work at their own pace. As you circulate, here are some things to notice.

   - In worksheet step 10, listen to students as they discuss the properties of the square they used in the construction. Students should make the following points.

     When we rotated the segment by 90 degrees, we were making right angles.

     We were making an image of the side, so the sides are congruent.

   - In worksheet step 20, students may say that a right angle has a right, or 90°, angle. ***How do you know it has a 90-degree angle?*** [Perpendicular lines form 90° angles.]
   - In worksheet steps 21 and 22, check to be sure that students label their shapes as shown in the illustration. Explain that this will match the way the Pythagorean Theorem is usually stated and will make it easier for students to remember it. Students can use the **Text** tool to reposition labels, if needed.

- In worksheet step 23, some students will have difficulty attaching the squares to the vertices of the triangle. Suggest that students try attaching the right vertex first and then the left vertex.
- In worksheet steps 27–29, listen to students as they work on these steps. If students are having difficulty seeing any relationships, suggest that they use the **Arrow** tool to drag the measurements so that the side lengths are listed vertically in order from side length *a* to side length *c*, and the area of a square is next to the length of its attached side. You can also ask some pointed questions. ***Do you see a relationship between the two smaller square areas and the largest square area? How is a side length related to the area of its square?***
- If students have time for the Explore More, they will investigate whether the converse of the Pythagorean Theorem is true. Have students try finding more than one triangle where the sum of the two areas is equal to the third area. ***Can you make a good conjecture based on one triangle? Can you find another triangle where the sum of the two areas is equal to the third area? What type of triangle is this one?***

6. If students will save their work, remind them where to save it now.

SUMMARIZE

Project the sketch. Expect to spend about 5 minutes.

7. Gather the class. Students should have their worksheets with them. Open **Squaring the Sides Present.gsp** and go to page "Pythagorean Theorem."
8. Begin the discussion by reviewing the definition of a right triangle. ***What is a right triangle?*** Write "A right triangle is" on chart paper. Work with students to write a definition. Here is a sample definition: A right triangle is a triangle that has one right angle, or a 90° angle. ***How do you know the triangle you constructed was a right triangle?*** [It was constructed with perpendicular sides; perpendicular lines form right angles.]
9. ***What is the side opposite the right angle called?*** [Hypotenuse] ***What are the other sides called?*** [Legs] ***In the sketch, which side is the hypotenuse and which sides are the legs?*** [The hypotenuse is side *c*; the legs are sides *a* and *b*.] Drag the right triangle around so students can see the right triangle in many positions and in many sizes. Point out that

the longest side is always the hypotenuse and the right angle is always opposite it.

10. Review how to find the area of a square. ***What is the formula for area of a square?*** [$A = s \times s$, or s^2] ***How would you find the area of any size red square in the sketch?*** [$A = b \times b$, or b^2] Have students state how they would find the areas of the other two squares as well.

11. Drag the vertices of the triangle. ***Did you find a relationship among the three areas? Explain.*** Students may make the following statements.

    Yes, we noticed that the sum of the areas of the two smaller squares equals the area of the largest square.

    We found the same thing, but we said it differently. The area of the largest square minus the area of the smallest square equals the area of the mid-sized square.

    Let's check your reasoning. Press *Show Calculation.* Drag the triangle vertices again to show that no matter the size of the right triangle, the sum of the areas of the two smaller squares equals the area of the largest square.

12. ***Based on your observations, what equation did you write that relates side lengths a, b, and c in any right triangle? Explain your thinking.*** Students should reason that the area of the square with side length a is a^2; the area of the square with side length b is b^2; and the area of the square with side length c is c^2. The sum of the areas of the two smaller squares equals the area of the largest square, or $a^2 + b^2 = c^2$. Tell students this is known as the Pythagorean Theorem and it holds true for any right triangle with leg lengths a and b and hypotenuse length c.

13. If time permits, discuss the Explore More. ***What did you discover?*** Students should find that the converse of the Pythagorean Theorem holds true: If the area of the largest square is equal to the sum of the areas of the smaller squares, then the triangle is a right triangle.

14. ***A right triangle has legs with lengths of 6 cm and 8 cm. How could you find the length of the hypotenuse?*** [Take the square root of the sum of the squares of the sides: $6^2 + 8^2 = c^2$; $\sqrt{6^2 + 8^2} = \sqrt{c^2}$; $\sqrt{36 + 64} = c$; $\sqrt{100} = c$; $10 = c$]

EXTEND

You might extend this activity by asking students whether they think the relationship among the areas holds for other similar shapes built along the sides of the right triangle. ***If you constructed equilateral triangles along the sides of the right triangle, do you think the area of the largest triangle would equal the sum of the areas of the smaller triangles?*** [The areas of the regions built on the legs of a right triangle will always total the area of the region built on the hypotenuse as long as all three regions are similar.] If you have time, you can model this fact.

ANSWERS

10. This construction uses the following property: A square has right angles and congruent sides.

20. This construction uses the following property: A right triangle has one side perpendicular to another side.

28. The sum of the areas of the two smaller squares equals the area of the largest square.

29. $a^2 + b^2 = c^2$

    Students already familiar with the Pythagorean Theorem are likely to write this when they make a conjecture. You might also have them express the theorem in words.

30. If the sum of the areas of squares on two sides of a triangle equals the area of the square on the third side, the triangle must be a right triangle.

Squaring the Sides

Name:

In this activity you'll construct squares on the sides of a right triangle using a custom tool. Then you'll use the areas of these squares to explore perhaps the most famous relationship in mathematics—the Pythagorean Theorem.

CONSTRUCT

1. In a new sketch, choose **Edit | Preferences.** On the Text panel, under **Show labels automatically,** check **For all new points.** Click **OK.**

In steps 2–9, you will create a custom tool that constructs a square.

2. Construct $\overline{AB}$.

3. You'll mark point *A* as a center and rotate point *B* and $\overline{AB}$ by 90°.
   Double-click point *A* to mark it as a center.
   Select point *B* and $\overline{AB}$. Then choose **Transform | Rotate.**
   In the Rotate dialog box, select **Fixed Angle** and enter 90 for degrees. Click **Rotate.**

4. Mark point *B'* as a center and rotate point *A* and $\overline{B'A}$ by 90°.

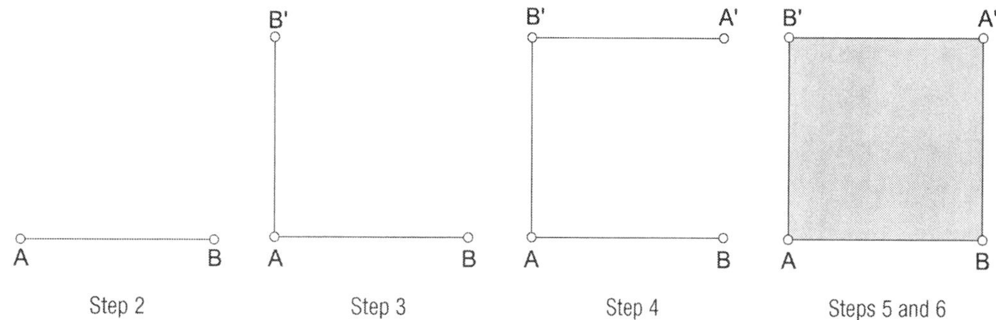

Step 2 Step 3 Step 4 Steps 5 and 6

5. Construct $\overline{A'B}$ to finish the square.

6. Select the vertices in order and choose **Construct | Quadrilateral Interior.**

7. Drag each vertex of the square to make sure it holds together.

8. Hide the labels by clicking on each point.

9. Now you'll make a custom tool of this construction that you can use to make other squares.

Select the entire figure. One way is to choose **Edit | Select All.**

Choose **Create New Tool** from the Custom Tools menu.

In the Tool Box dialog box, enter Square for the Tool Name.

EXPLORE

10. What properties of a square did you use in this construction?

11. Choose **Edit | Preferences.** On the Text panel, under **Show labels automatically,** uncheck **For all new points.** Click **OK.**

12. Use the custom tool to get a feel for the way it works. Note that the direction in which the square is constructed depends on how you use the tool.

CONSTRUCT

13. Now you'll add a new page to your sketch.

Choose **File | Document Options.**

In the Document Options dialog box, select Add Page and choose **Blank Page.** Click **OK.**

In steps 14–18, you'll construct a right triangle. Then in step 23 you'll use your **Square** tool to attach squares onto each side of the right triangle.

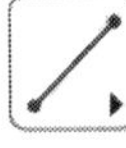

14. Construct $\overline{AB}$. Label the points if you find it helpful, but you'll be changing the labels in step 21.

15. Select point *A* and $\overline{AB}$ and choose **Construct | Perpendicular Line.**

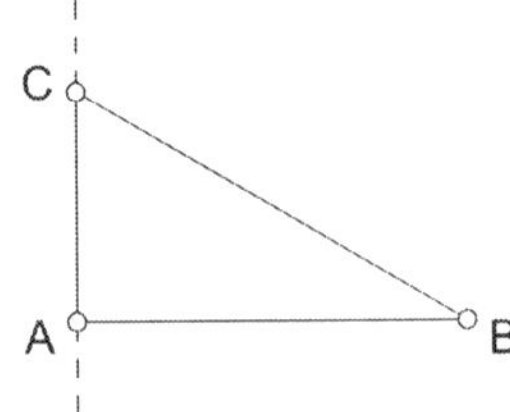

16. Construct point *C* on the perpendicular line.

17. Select $\overleftrightarrow{AC}$ and choose **Display | Hide Perpendicular Line.**

18. Construct $\overline{AC}$ and $\overline{BC}$.

19. Drag each vertex to confirm that your triangle stays a right triangle.

20. What property of a right triangle did you use in your construction?

21. Click on the vertices in order from *A* through *C* as shown at right, so that *C* is the right-angle vertex.

    To change a label, double-click it, enter the new label, and click **OK.**

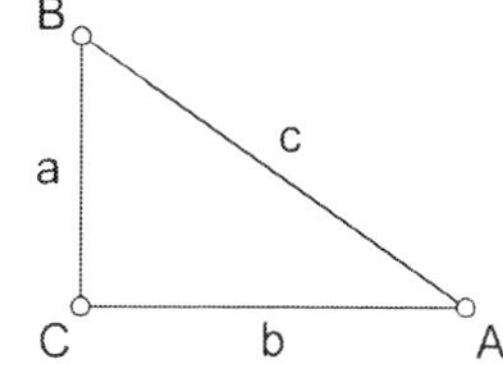

22. Show the labels of the sides by clicking on each segment. Change the labels to *a, b,* and *c* so that side *a* is opposite $\angle A$, side *b* is opposite $\angle B$, and side *c* is opposite $\angle C$.

23. Use your new **Square** custom tool to construct squares on the sides of your triangle. Be sure to attach each square to a pair of the triangle's vertices. If your square goes the wrong way (overlaps the interior of your triangle) or is not attached properly, choose **Edit | Undo** and try attaching the square to the triangle's vertices in the opposite order.

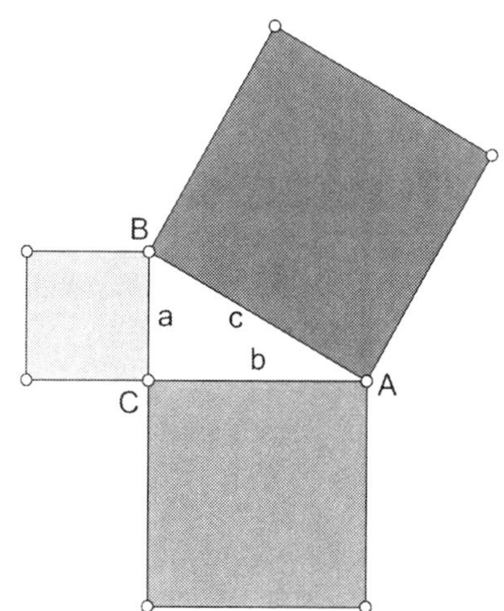

24. Drag the vertices of the triangle to make sure the squares are properly attached.

EXPLORE

25. Measure the areas of the three squares by selecting each interior and choosing **Measure | Area.**

26. Measure the lengths of sides *a, b,* and *c* by selecting each side and choosing **Measure | Length.**

27. Drag each vertex of the triangle and observe the measures.

28. Now you'll use the Calculator to write an expression based on your observations.

    Choose **Number | Calculate** to show the Sketchpad Calculator.

    Click once on a measurement to enter it into a calculation.

    Describe any relationship you see among the three areas.

29. Based on your observations about the areas of the squares, write an equation that relates *a, b,* and *c* in any right triangle. (*Hint:* What's the area of the square with side length *c*? What are the areas of the squares with side lengths *a* and *b*? How are these areas related?)

EXPLORE MORE

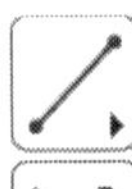

30. Now you'll investigate the converse of the Pythagorean Theorem.

    Construct a generic triangle (not a right triangle).

    Construct a square on each side.

    Measure the areas of the squares and find the sum of two of them. Drag until the sum is equal to the third area. What kind of triangle do you have?

3

Polygon Properties

Four-Sided Family: Defining Special Quadrilaterals

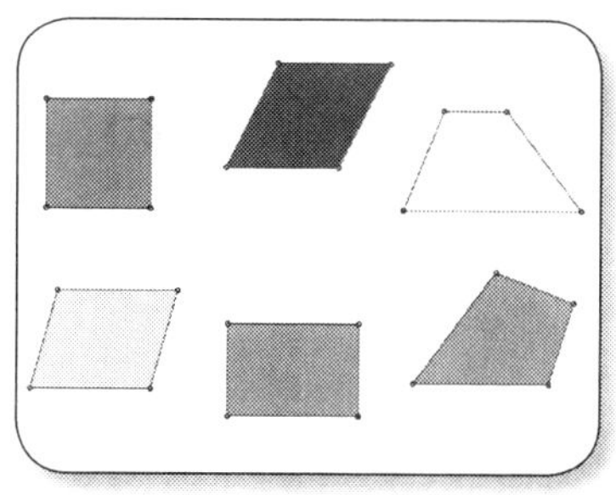

Using a prepared sketch, students explore properties of special quadrilaterals, including trapezoids, kites, parallelograms, rectangles, rhombuses, and squares. Students write definitions for each based on their observations and describe how the quadrilaterals are related.

Meet the Parallelogram: Properties of Parallelograms

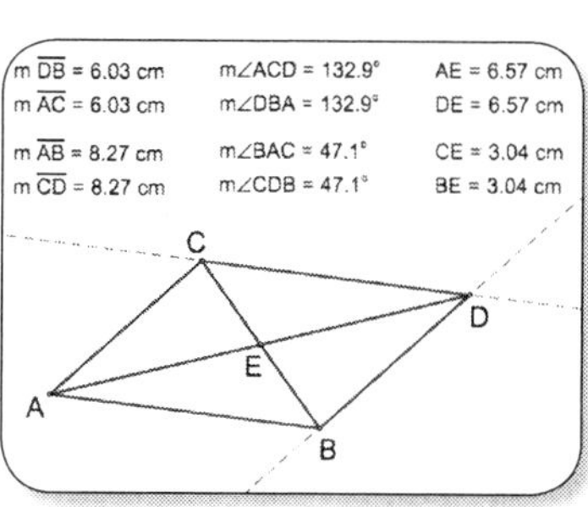

Students construct a parallelogram, measure side lengths and angles, and observe that opposite sides are congruent, opposite angles are congruent, and consecutive angles are supplementary. Then they construct the diagonals, measure the distances from the vertices to the point of intersection, and discover that the diagonals bisect each other.

Meet the Rectangle: Properties of Rectangles

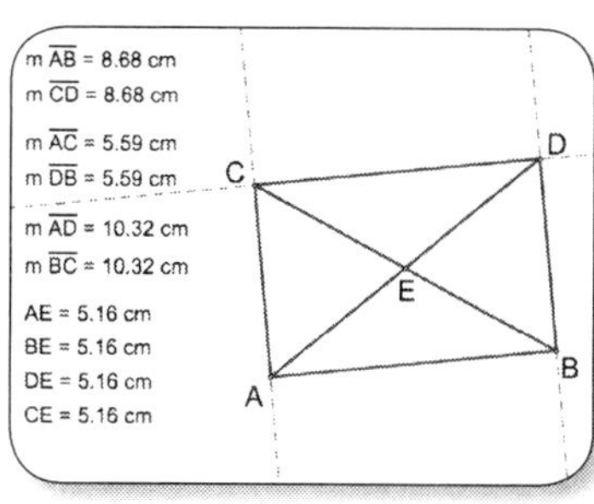

Students construct a rectangle, measure side lengths, and observe that opposite sides are congruent. Then they construct the diagonals, measure the distances from the vertices to the point of intersection, and discover that the diagonals are congruent and bisect each other.

Meet the Rhombus: Properties of Rhombuses

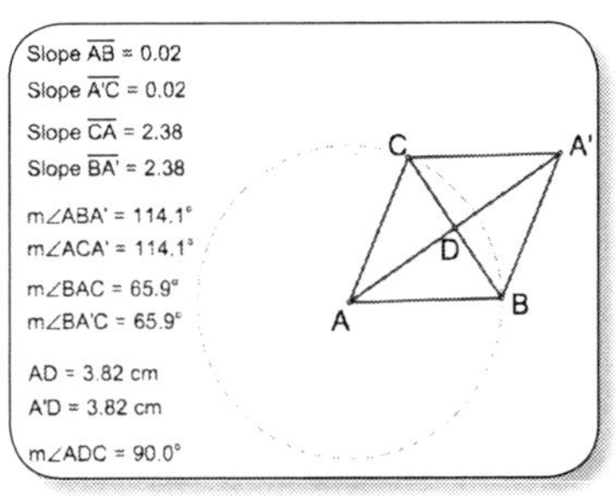

Students construct a rhombus, measure the angles and slopes of the sides, and observe that opposite sides are parallel, opposite angles are congruent, and consecutive angles are supplementary. Then they construct the diagonals, measure the distances from the vertices to the point of intersection and the angles formed, and discover that the diagonals are perpendicular bisectors of each other and bisectors of the interior angles.

Quadrilateral Pretenders: Classifying Quadrilaterals

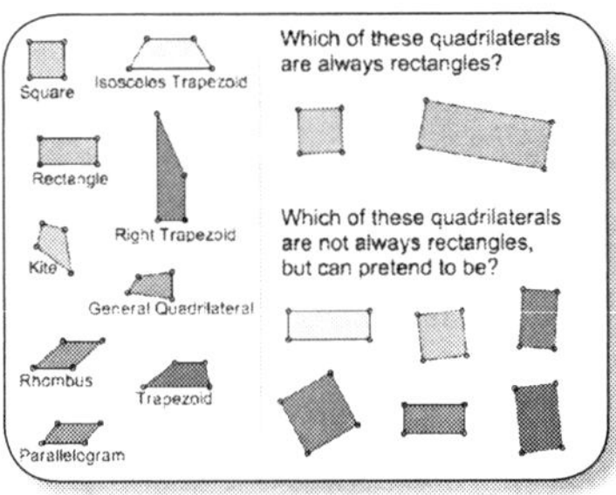

Students drag edges and vertices of various quadrilaterals to discover which are constructed to have specific characteristics. As they make distinctions on the basis of characteristics, they deepen their understanding of the definitions of various quadrilaterals, their properties, and the relationships among them.

All Things Being Equal: Constructing Regular Polygons

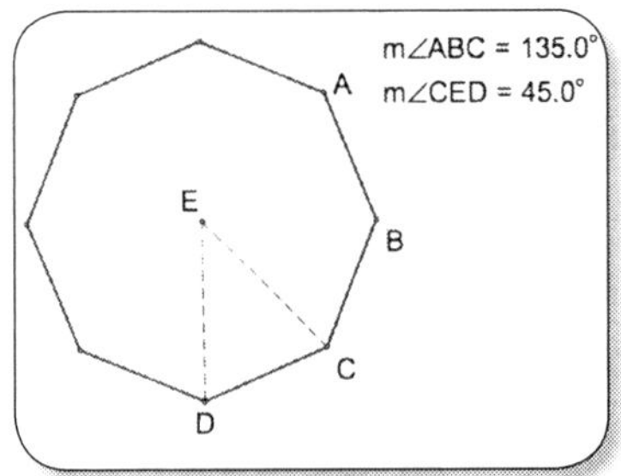

Students construct regular polygons using rotations. First they use the central angle as the angle of rotation, then the interior angles.

Polygon Pretenders: Classifying Polygons

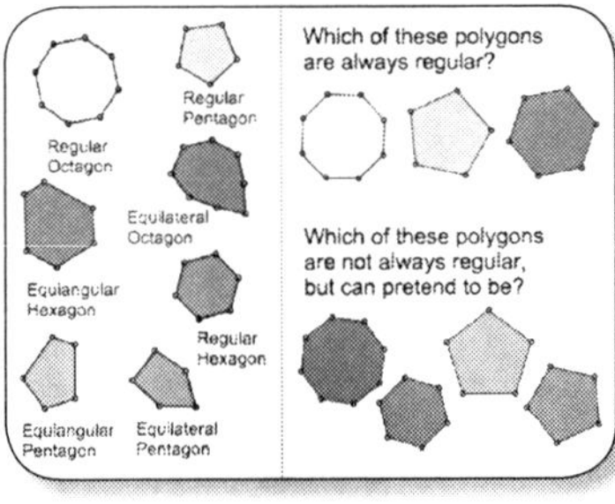

Students drag vertices of various polygons to discover which are constructed to have specific characteristics. As they make distinctions on the basis of characteristics, they understand the definitions of various classifications of polygons, their properties, and the relationships among them.

Four-Sided Family: Defining Special Quadrilaterals

ACTIVITY NOTES

INTRODUCE

Project the sketch for viewing by the class. Expect to spend about 10 minutes.

1. Open **Four-Sided Family.gsp** and go to page "Special Quads." Enlarge the document window so it fills most of the screen.
2. Explain, ***Today you're going to use Sketchpad to explore the properties of some special quadrilaterals. What is a quadrilateral?*** Work with students to come up with a definition. Write it on chart paper. Here is a sample definition: *A quadrilateral is a four-sided polygon.*
3. ***What are some examples of quadrilaterals in the real world?*** Students may make the following observations.

   The cover of our math book is a quadrilateral.

   The window has four sides, so it is a quadrilateral.

   The top of my desk is a quadrilateral.

   The tiles on the floor are quadrilaterals that are all the same size and shape.

   Encourage students to find some quadrilaterals that do not have four right angles, such as kites or trapezoidal desks.
4. ***There are some quadrilaterals that have special properties: square, rectangle, rhombus, parallelogram, trapezoid, and kite. You may know about some of these quadrilaterals already; others may be unfamiliar to you. You're going to investigate these quadrilaterals using Sketchpad, and then we'll organize your findings in a chart.***
5. ***Before you begin I'll demonstrate how to use the sketch.*** Model worksheet steps 1–3. Here are some tips.
   - In worksheet step 2, model how to drag various parts of the trapezoid. ***Each quadrilateral has a different set of constraints in its construction that defines it. For example, you can drag vertices of the trapezoid to make different shapes, but you'll still have a trapezoid.***
   - Tell students that they can navigate between pages by clicking the tabs along the bottom of the sketch.
6. If you want students to save their work, demonstrate choosing **File | Save As,** and let them know how to name and where to save their files.

DEVELOP

Expect students at computers to spend about 35 minutes.

7. Assign students to computers and tell them where to locate **Four-Sided Family.gsp.** Distribute the worksheet. Tell students to work through step 19 and do the Explore More if they have time. Encourage students to ask their neighbors for help if they are having difficulty with Sketchpad.

8. Let pairs work at their own pace. As you circulate, here are some things to notice.

   - In worksheet step 3, define *slope* for students, if needed. ***A slope is a measure of the steepness of a line or a segment.*** For this activity students need to know only that lines, or segments, with the same slope are parallel.

   - When students drag the quadrilaterals to observe certain properties, encourage them to make several different quadrilaterals each time to test any conjectures they may make during their observations. ***As you drag parts of your quadrilateral, ask yourself these questions each time: How are the lengths of the sides related? Are any sides parallel? How are the measures of the angles related?***

   - In worksheet step 10, be sure students select the vertex as the second point when measuring an angle. ***The vertex of your angle must be the second point you select when defining your angle.*** Review with students, as needed, how angles are named by three points.

   - When students describe the angles in worksheet steps 11 and 15, encourage them to name the type of angle. ***What's another name for a 90-degree angle?*** [Right angle] Students may recognize that the opposite sides must be parallel as well because the angles are all right angles forming perpendicular line segments.

   - In worksheet steps 16 and 17, students discover that a square is both a rhombus and a rectangle. ***When is the rhombus also a rectangle?*** [When it has a right angle, or when it's a square] ***When is the rectangle also a rhombus?*** [When its sides are all congruent, or when it's a square]

   - In worksheet step 19, the statements test students' understanding of the properties of the quadrilaterals and how they are all related. Listen to students as they work through these statements. Encourage them to test their answers by dragging the quadrilaterals in the sketch.

- If students have time for the Explore More, they will use their definition of a rectangle to construct a rectangle that keeps its defining properties when it is dragged. ***What is your definition of a rectangle, based on your exploration today? How can you construct a rectangle that has four right angles? What types of lines will you need to construct?*** [Perpendicular lines] If students need a hint, have them choose **Display | Show All Hidden** to see how the rectangle in the sketch was constructed.

9. If students will save their work, remind them where to save it now.

SUMMARIZE

Project the sketch. Expect to spend about 15 minutes.

10. Gather the class. Students should have their worksheets with them. Begin the discussion by opening **Four-Sided Family.gsp.**

11. ***Let's review what you learned about these special quadrilaterals.*** Make the following table on chart paper, leaving the definitions blank. Then have volunteers come up and use the sketch to support their definitions. Fill in the table as you go over each quadrilateral.

Quadrilateral	Definition
Trapezoid	one pair of parallel sides
Kite	two pairs of congruent adjacent sides
Parallelogram	two pairs of parallel opposite sides
Rectangle	four right angles
Rhombus	four congruent sides
Square	four congruent sides and four right angles

12. Some texts define trapezoids as having *at least* one pair of parallel sides, whereas others define them as having *exactly* one pair of parallel sides. ***Which definition is being used in this sketch?*** Because you can drag the trapezoid in the sketch to form a parallelogram, the "at least" definition might be more appropriate here. You might want to discuss the differences between these definitions.

13. For the kite definition, students may say "next to" instead of "adjacent." Introduce this term as needed. **Adjacent sides** ***means that two sides share a common vertex.***

14. Review students' answers to worksheet steps 16–18. ***How did you drag the rhombus so that it was a rectangle?*** [Dragged it until four right angles formed] ***What did it look like?*** [Square] ***How did you drag the rectangle so that it was a rhombus?*** [Dragged it until it had four congruent sides] ***What did it look like?*** [Square] ***How did you rewrite your definition of square?*** Students should understand that a square is also a rhombus and a rectangle.

15. For the answers to worksheet step 19, have volunteers use the sketch to support each statement. For example, students should drag the parallelogram so that it is and is not a square; it is *sometimes* a square.

16. If time permits, discuss the Explore More. Have students demonstrate how they constructed the rectangle and have them drag parts of it to prove that it remains a rectangle.

17. ***Make a graphic that shows the relationships among the different types of quadrilaterals you investigated today.*** You may wish to have students respond individually in writing to this prompt. Students may use a chart or possibly a Venn diagram to show that the trapezoid, kite, parallelogram, rectangle, rhombus, and square are all quadrilaterals; the rectangle, rhombus, and square are all parallelograms; and the square is both a rectangle and a rhombus.

EXTEND

What questions occurred to you while you were exploring quadrilaterals? Encourage curiosity. Here are some sample student queries.

Is a trapezoid with one pair of congruent opposite sides called something special?

Do kites also have a pair of congruent angles?

Can you have different definitions of the same kind of figure?

Are there kinds of quadrilaterals other than these?

Is there anything special about the diagonals of these quadrilaterals?

Are there relationships like these among shapes other than quadrilaterals?

ANSWERS

5. A trapezoid is a quadrilateral with one pair of parallel sides.
7. A kite is a quadrilateral with two pairs of congruent adjacent sides.
9. A parallelogram is a quadrilateral with two pairs of parallel opposite sides.
11. A rectangle is a quadrilateral with four congruent (or right) angles.
13. A rhombus is a quadrilateral with four congruent sides.
15. A square is a quadrilateral with four congruent sides and four congruent angles.
16. A square
17. A square
18. A square is a quadrilateral that is both a rhombus and a rectangle.
19. a. Sometimes b. Sometimes c. Always
    d. Always e. Never
20. Red is a trapezoid; blue is a kite; yellow is a rectangle; green is a rhombus; purple is a square; brown is a parallelogram.
21. Constructions may vary. Construct $\overline{AB}$. Then construct lines perpendicular to $\overline{AB}$ through point A and point B. Construct point C on the line through point A. Construct a line through point C perpendicular to $\overleftrightarrow{AC}$. Construct a point D at the intersection of the line through point C and point B. Hide the lines, and then construct segments to complete the rectangle.

Four-Sided Family

Name:

Trapezoids, kites, parallelograms, rectangles, rhombuses, and squares have special properties that distinguish them from other quadrilaterals. Explore what makes them different from ordinary quadrilaterals.

EXPLORE

1. Open **Four-Sided Family.gsp** and go to page "Special Quads."

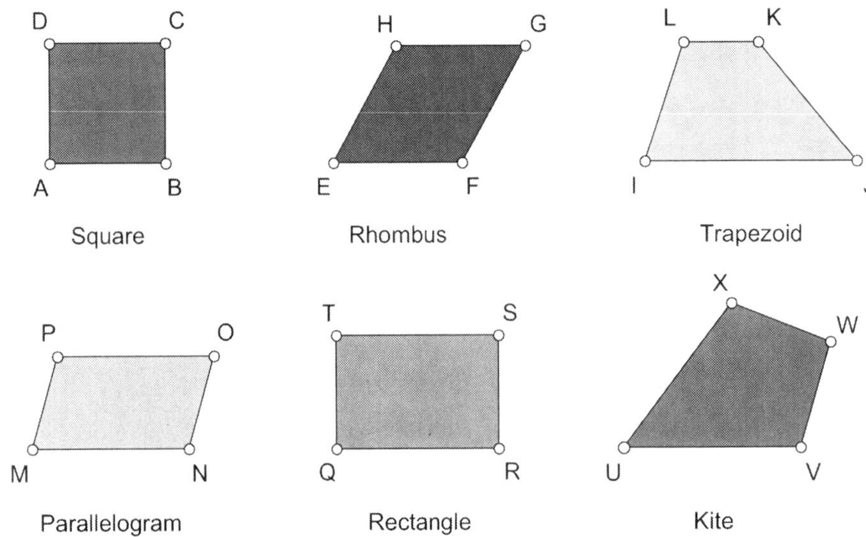

2. Drag various parts of the quadrilaterals to see how they behave.

3. Go to page "Trapezoid." Select the four sides of the trapezoid and choose **Measure | Slope.**

4. If necessary, click on sides of the trapezoid to label them, or double-click on labels to change them.

5. Drag parts of the trapezoid. If lines (or segments) have the same slope, they are parallel. How many pairs of sides in the trapezoid are always parallel? Use your observations to define *trapezoid.*

6. Go to page "Kite." Select the four sides of the kite and choose **Measure | Length.**

7. Drag parts of the kite. Which sides are always equal in length? Use your observations to define *kite.*

8. Go to page "Parallelogram." Measure the slopes of all four sides of the parallelogram.

9. Drag parts of the parallelogram. How many pairs of sides are parallel? Use your observations to define *parallelogram.*

10. Go to page "Rectangle." Measure the angles in the rectangle. To measure an angle, select three points with the vertex as your second point, and choose **Measure | Angle.**

11. Drag parts of the rectangle. Use your observations to define *rectangle.*

12. Go to page "Rhombus." Measure the side lengths of the rhombus.

13. Drag parts of the rhombus. Use your observations to define *rhombus.*

14. Go to page "Square." Measure the side lengths and angles of the square.

15. Drag parts of the square. Use your observations to define *square.*

Four-Sided Family

continued

16. Go back to page "Special Quads." Drag the rhombus so that it's also a rectangle (or at least close to it). What's the best name for this shape?

17. Drag the rectangle so that it's also a rhombus (or at least close to it). What's the best name for this shape?

18. Based on your observations in steps 16 and 17, write a definition of *square* different from your definition in step 15.

19. Circle the word—*always, sometimes,* or *never*—that makes each of the following sentences true.
    a. A parallelogram is (always/sometimes/never) a square.
    b. A rectangle is (always/sometimes/never) a rhombus.
    c. A square is (always/sometimes/never) a rhombus.
    d. A rectangle is (always/sometimes/never) a parallelogram.
    e. A parallelogram that is not a rectangle is (always/sometimes/never) a square.

EXPLORE MORE

20. Go to page "Explore More." Drag the vertices and observe each shape's behavior. What type of quadrilateral is each shape?

21. Open a new sketch by choosing **File | New Sketch.** Construct a rectangle using the definition you wrote today. Make sure your rectangle keeps its defining properties when you drag parts of it. Describe your construction method.

Meet the Parallelogram: Properties of Parallelograms

INTRODUCE

Project the sketch for viewing by the class. Expect to spend about 10 minutes.

1. Open Sketchpad and enlarge the document window so it fills most of the screen.

2. Explain, ***Today you're going to use Sketchpad to construct a parallelogram and investigate its properties. First I'll demonstrate how to construct a parallelogram using its definition. As I construct the parallelogram, think about how you would write the definition.*** As you demonstrate, make lines thick and labels large for visibility. Model the parallelogram construction in worksheet steps 1–12. Here are some tips.

   - In worksheet steps 1–5, you will construct one side of the parallelogram and a line parallel to it. ***What does it mean for two lines to be parallel?*** [The lines never intersect.] ***What does $\overline{AB}$ represent?*** [One side of the parallelogram] ***What do you think point C is?*** [A vertex]

   - In worksheet step 6, you will construct a second side that is adjacent to $\overline{AB}$. ***What does $\overline{AC}$ represent?*** [Another side of the parallelogram] ***What is its relationship to $\overline{AB}$?*** [It is next to, or adjacent to, it.]

   - In worksheet steps 8 and 9, model how to use the **Point** tool to construct the point of intersection (although you can also construct an intersection point by clicking the intersection with the **Arrow** tool). Explain that both lines will be highlighted when the **Point** tool is in the right place to construct the point of intersection. ***What is point D?*** [A vertex]

   - In worksheet steps 10 and 11, explain that two sides of the parallelogram are lines, so you will hide them and construct segments instead. Ask students to name the finished parallelogram. [Parallelogram *ABDC* or *ACDB*] Review that parallelograms are named by their vertices. Depending on your curriculum, you might introduce the symbol for parallelogram: ▱.

   - In worksheet step 12, demonstrate how to drag the different parts of the parallelogram to test the construction. Explain that this is an important step; if the parallelogram is not constructed properly, students can choose **Edit | Undo** as needed.

3. ***How does this construction help you define* parallelogram*?*** Students may make the following response.

   Based on the construction, we know side AB is parallel to side CD and that side AC is parallel to side BD. The sides are not adjacent, but are opposite. So a definition would be that opposite sides are parallel.

4. Encourage students to add to the definition if it is not complete. ***How many sides does a parallelogram have?*** [Four sides] ***What type of polygon is it?*** [Quadrilateral]

5. After students have come up with a complete definition, write it on chart paper: *A parallelogram is a quadrilateral whose opposite sides are parallel.*

6. If you want students to save their work, demonstrate choosing **File | Save As,** and let them know how to name and where to save their files.

DEVELOP

Expect students at computers to spend about 25 minutes.

7. Assign students to computers. Distribute the worksheet. Tell students to work through step 21 and do the Explore More if they have time. Encourage students to ask their neighbors for help if they are having difficulty with the construction.

8. Let pairs work at their own pace. As you circulate, here are some things to notice.

   - In worksheet step 13, ask students what they observe about the lengths of the sides. Students should notice that opposite sides are congruent. ***Do you think this holds true for any parallelogram?*** Have students think about this before dragging the parallelogram and making their conjectures.

   - In worksheet step 14, be sure students choose the vertex as the second point when measuring an angle. Also, check that students don't measure the same angle twice. Encourage students to look for relationships among the angle measures. ***What do you notice about angles opposite each other? How about the consecutive angles, the angles next to each other?*** Students should observe that opposite angles have the same measure and adjacent angles are supplementary.

Review the term *supplementary*, if needed. ***What are supplementary angles?*** [Angles whose measures add to 180°.] ***Can you find any of these types of angles?***

- In worksheet step 15, as students drag parts of the parallelogram, encourage them to make different types of parallelograms. ***Now make another parallelogram that looks completely different. What changes as you drag parts of the parallelogram? What stays the same?***
- In worksheet step 17, check that students understand that a diagonal is a line segment that connects two nonadjacent vertices. ***How will you construct the diagonals of the parallelogram?*** [Construct $\overline{AD}$ and $\overline{CB}$]
- In worksheet step 20, students will be measuring distances between points, although they could also construct new segments from the vertices to the point of intersection and then measure their lengths. Again, remind students to drag the parts of the parallelogram to form many different parallelograms. ***What do you notice about the distances you measured?***
- If students have time for the Explore More, they will use the properties of a parallelogram to construct parallelograms using different methods. Be sure students clearly describe their construction methods and explain what properties they used. The more time you give students, the more methods they will find. Give students hints to start their thinking, if necessary. Have students drag vertices of their figures to make sure their constructions are correct. Parallelograms that fall apart and can turn into other shapes are underconstrained. A construction that stays a parallelogram but that can't take on all possible shapes of a parallelogram is overconstrained.

9. If students will save their work, remind them where to save it now.

SUMMARIZE

Project the sketch. Expect to spend about 10 minutes.

10. Gather the class. Open **Meet Parallelogram Present.gsp** and use pages "Sides and Angles" and "Diagonals" as needed. Students should have their worksheets with them. Begin the discussion by writing "Properties of Parallelograms" on chart paper. ***We started this activity by constructing a parallelogram based on its definition. What property did we use to construct the parallelogram?*** Write down the response

on the chart paper. Then review the conjectures students made about the sides, the angles, and diagonals of parallelograms. Volunteers may come to the computer and drag the parallelogram to show how the conjectures hold true for different parallelograms.

Properties of Parallelograms
Opposite sides are parallel.
Opposite sides are equal in length.
Opposite angles have equal measures.
Consecutive angles are supplementary.
Diagonals bisect each other.

11. Drag the parallelogram so that it is a rectangle, having four right angles. ***What shape does this look like?*** [Rectangle] ***Is it still a parallelogram? Explain.*** [Yes, it still has the properties of a parallelogram.]
12. Drag the parallelogram so that it is a rhombus, having four congruent sides. ***What shape does this look like?*** [Rhombus] ***Is it still a parallelogram? Explain.*** [Yes, it still has the properties of a parallelogram.]
13. Now drag the parallelogram so that it is a square, having four right angles and four congruent sides. It might be easier if you select one point at a time and use the arrow keys on your computer to move it in small increments. ***What shape does this look like?*** [Square] ***Is it still a parallelogram? Explain.*** [Yes, it still has the properties of a parallelogram.]
14. If time permits, discuss the Explore More. Have students review their construction methods and the properties they used in each. The last four pages of **Meet Parallelogram Present.gsp** show some methods that you may wish to share with students.
15. ***The measure of one angle of a parallelogram is 60 degrees. What are the measures of the other angles? Explain your reasoning.*** [Opposite angles are congruent, so the opposite angle measures 60°. Consecutive angles are supplementary, so the two other angles each measure 120°.]

EXTEND

1. Have students draw a Venn diagram showing the relationship between quadrilaterals, parallelograms, rectangles, rhombuses, and squares.

2. Have students explore what conditions are necessary for a quadrilateral to be a parallelogram. Ask students to construct a quadrilateral where the given condition is true and then state whether the quadrilateral is always, sometimes, or never a parallelogram.

Condition	Always, Sometimes, Never a Parallelogram
The diagonals bisect each other.	Always
The diagonals are congruent.	Sometimes (when the parallelogram is a square diagonals bisect each other, making a rectangle)
Opposite angles are 45° and 50°.	Never
Both pairs of opposite sides are congruent.	Always

3. Ask students to try to construct a parallelogram with all its vertices on the circumference of a circle. ***Were you able to construct a parallelogram? If so, what type of parallelogram did you construct?*** [Rectangle]

ANSWERS

16. Opposite sides in a parallelogram are equal in length. Opposite angles in a parallelogram are equal in measure. Consecutive angles in a parallelogram are supplementary.

21. Diagonals in a parallelogram bisect each other. Students might also notice that the diagonals in a parallelogram divide the parallelogram into four pairs of congruent triangles.

22. Students may find one or more of the following methods.

Method: Construct a segment and a point not on the segment. Mark the endpoints of the segment as a vector. Translate the point not on the segment by the marked vector. Construct the missing sides.
Properties: A parallelogram has one pair of opposite sides that are both parallel and congruent.

Method: Construct a segment and a point not on the segment. Select the segment and the point and choose **Construct | Circle by Center+Radius**. Construct a line through the point, parallel to the segment. Find the point where the circle and the line intersect. This is the fourth vertex of the parallelogram.

Properties: A parallelogram has one pair of opposite sides that are both parallel and congruent.

Method: Construct a segment and its midpoint. Construct a line through the segment attached at the midpoint. Construct a circle centered at the midpoint. Find the two points where the circle intersects the line. These two points and the original endpoints of the segment are the vertices of the parallelogram.
Properties: The diagonals of a parallelogram bisect each other.

Method: Construct a pair of concentric circles. Construct two lines that both contain the center point. Find the two points of intersection of one of the lines with one of the circles and of the other line with the other circle. These four points are the points of intersection of the parallelogram.
Properties: The diagonals of a parallelogram bisect each other.

Meet the Parallelogram

Name:

In this activity you'll learn how to construct a *parallelogram* (a quadrilateral whose opposite sides are parallel). Then you'll discover some properties of parallelograms.

CONSTRUCT

1. In a new sketch, construct a segment.

2. Label the endpoints *A* and *B*.

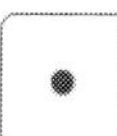

3. Construct a point above $\overline{AB}$.

4. Label this point *C*.

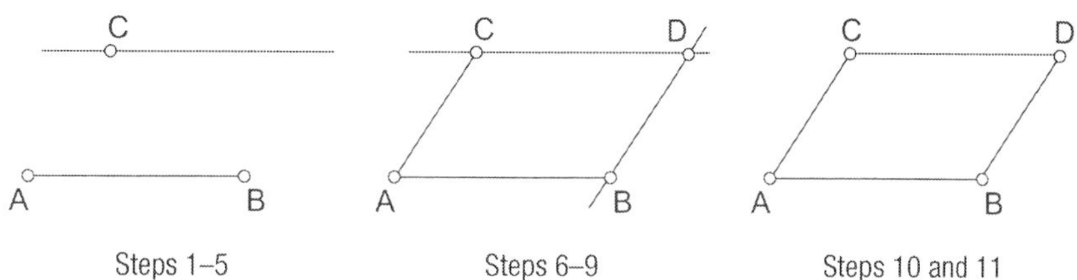

5. Now you'll construct a line through point *C* parallel to $\overline{AB}$.
   Select $\overline{AB}$ and point *C*, and then choose **Construct | Parallel Line.**

6. Construct $\overline{AC}$.

7. Construct a line through point *B* parallel to $\overline{AC}$.

8. Construct a point where the two lines intersect.

9. Label this point *D*.

10. Select both lines and choose **Display | Hide Parallel Lines.**

11. Construct the missing segments $\overline{CD}$ and $\overline{BD}$.

12. Drag different vertices of your parallelogram to make sure it's constructed properly.

EXPLORE

13. Measure the sides by selecting them and choosing **Measure | Length.**

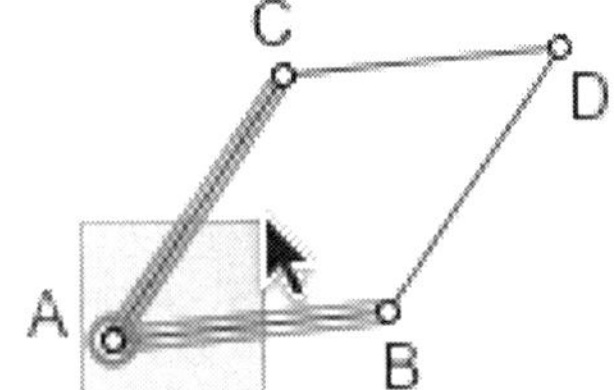

14. Now measure the angles.

    To measure an angle, select three points with the vertex as the second point. Press and drag to make a selection rectangle as shown. Then choose **Measure | Angle.**

15. Drag different parts of the parallelogram and observe the measurements.

16. Write at least three conjectures about the sides and angles of a parallelogram.

CONSTRUCT

17. Construct the diagonals of the parallelogram.

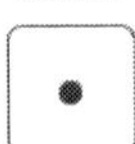

18. Construct the point of intersection of the diagonals.

19. Label it point *E*.

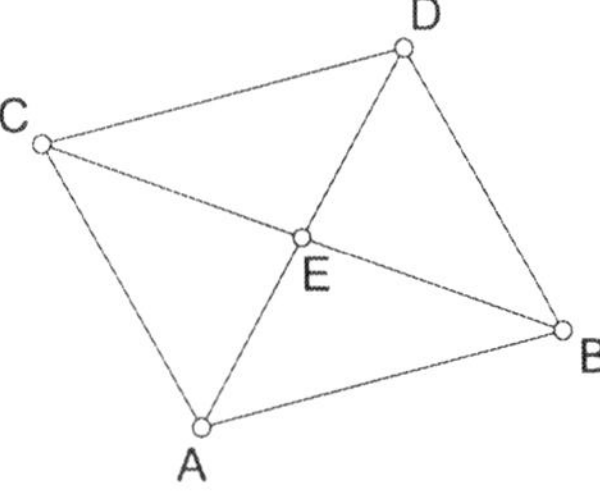

EXPLORE

20. Drag parts of the parallelogram and observe the diagonals. Measure lengths that look as though they might be related.

    To measure a distance between two points, select the two points and then choose **Measure | Distance.**

21. Write a conjecture about the diagonals of a parallelogram.

EXPLORE MORE

22. How many ways can you construct a parallelogram? Try methods that use the Construct menu, the Transform menu, or combinations of both. Consider how you might use diagonals. Write a brief description of each construction method along with the properties of parallelograms that make that method work.

Meet the Rectangle: Properties of Rectangles

ACTIVITY NOTES

INTRODUCE

Project the sketch for viewing by the class. Expect to spend about 10 minutes.

1. Open Sketchpad and enlarge the document window so it fills most of the screen.
2. Explain, ***Today you're going to use Sketchpad to construct a rectangle and investigate its properties. First I'll demonstrate how to construct a rectangle using its definition. As I construct the rectangle, think about how you would write the definition.*** As you demonstrate, make lines thick and labels large for visibility. Model the rectangle construction in worksheet steps 1–11. Here are some tips.
   - In worksheet steps 1–3, you will construct one side of the rectangle and two lines perpendicular to it. ***What does it mean to say that two lines are perpendicular?*** [The lines intersect at right angles.] ***What does*** $\overline{AB}$ ***represent?*** [One side of the rectangle]
   - In worksheet steps 4–8, you'll construct and label point *C* on the line through point *A*. Next you'll construct a line perpendicular to $\overleftrightarrow{AC}$. Then you'll construct and label point *D*, the point of intersection of this line and the line through point *B*. Model how to use the **Point** tool to construct the point of intersection (although you can also construct an intersection point by clicking the intersection with the **Arrow** tool). Explain that both lines will be highlighted when the **Point** tool is in the right place to construct the point of intersection. ***What is point D?*** [A vertex]
   - ***What can you say about the relationship between*** $\overleftrightarrow{CD}$ ***and*** $\overleftrightarrow{DB}$***? How do you know?*** Help students understand that because both $\overleftrightarrow{AC}$ and $\overleftrightarrow{DB}$ are perpendicular to $\overline{AB}$, and $\overleftrightarrow{AC}$ is perpendicular to $\overleftrightarrow{CD}$, then $\overleftrightarrow{DB}$ is also perpendicular to $\overleftrightarrow{CD}$.
   - In worksheet steps 9 and 10, explain that three of the sides of the rectangle are lines, so you will hide them and construct segments instead. Model how to select each line (not the points) and choose **Display | Hide Perpendicular Lines.** ***How can we name this rectangle?*** [Rectangle *ACDB*] Review that rectangles are named by their vertices. Depending on your curriculum, you might want to introduce the symbol for rectangle: ▭
   - Demonstrate how to drag the different parts of the rectangle to test the construction. Explain that this is an important step.

If students make errors during the construction, remind them to use **Edit | Undo** as needed.

3. ***How does this construction help you define* rectangle?** Students may make the following responses.

   A rectangle is a shape with perpendicular sides.

   Based on the construction, we know that side AB and side CD are perpendicular to side CA and side DB. Perpendicular lines form right angles, so all four angles are 90 degrees. A rectangle has four 90-degree angles.

   Can we just say a rectangle has four right angles?

4. Encourage students to add to the definition if it is not complete. ***How many sides does a rectangle have?*** [Four sides] ***What type of polygon is it?*** [Quadrilateral] Students may say that a rectangle is also a parallelogram because it has opposite sides that are congruent and parallel. Stating that a rectangle is a quadrilateral is sufficient for the definition.

5. After students have come up with a complete definition, write it on chart paper: *A rectangle is a quadrilateral with four right angles.*

6. If you want students to save their work, demonstrate choosing **File | Save As,** and let them know how to name and where to save their files.

DEVELOP

Expect students at computers to spend about 25 minutes.

7. Assign students to computers. Distribute the worksheet. Tell students to work through step 19 and do the Explore More if they have time. Encourage students to ask their neighbors for help if they are having difficulty with the construction.

8. Let pairs work at their own pace. As you circulate, here are some things to notice.

   - Throughout, emphasize that students will need to notice when objects are highlighted or selected. If students make errors, remind them to choose **Edit | Undo** and try again.
   - In worksheet step 12, ask students what they observe about the lengths of the sides. They should notice that opposite sides are

congruent. ***Do you think this holds true for any rectangle?*** Have students think about this before dragging the rectangle and making their conjectures.

- In worksheet step 13, as students drag parts of the rectangle, encourage them to make different types of rectangles. ***Now make another rectangle that looks completely different. What changes as you drag parts of the rectangle? What stays the same?***
- In worksheet step 15, check that students understand that a diagonal is a line segment that connects two nonadjacent vertices. ***How will you construct the diagonals of the rectangle?*** [Construct $\overline{AD}$ and $\overline{CB}$]
- In worksheet step 16, again remind students that both diagonals will be highlighted when the point of intersection is chosen correctly.
- In worksheet step 18, students will be measuring distances between points, although they could also construct new segments from the vertices to the point of intersection and then measure their lengths. Again, remind students to drag the parts of the rectangle to form many different rectangles. ***What do you notice about the lengths or distances you measured?***
- If students have time for the Explore More, they will use the properties of a rectangle to construct a rectangle using a different method. Be sure students clearly describe their construction methods and explain what properties they used. Give students hints to start their thinking, if necessary. Have students drag vertices of their figures to make sure their constructions hold. Rectangles that fall apart and can turn into other shapes are underconstrained. A construction that stays a rectangle but that can't take on all possible shapes of a rectangle is overconstrained.

9. If students will save their work, remind them where to save it now.

SUMMARIZE

Project the sketch. Expect to spend about 15 minutes.

10. Gather the class. Open **Meet the Rectangle Present.gsp** and use pages "Sides" and "Diagonals" as needed. Students should have their worksheets with them. Begin the discussion by writing "Properties of Rectangles" on chart paper. ***We started this activity by constructing a rectangle based on its definition. What property did we use to construct the rectangle?*** Write down the response on the chart paper.

Then review the conjectures students made about the sides and the diagonals of rectangles. Volunteers may come to the computer and drag the rectangle to show how the conjectures hold true for different rectangles.

Depending on the background of your students, you might point out that $\overline{AB}$ can be thought of as a transversal that intersects $\overline{AC}$ and $\overline{BD}$. If students look at interior angles, $\angle CAB$ and $\angle DBA$, they will find that the angles are supplementary ($90° + 90° = 180°$), and when the interior angles are supplementary, the transversal intersects parallel lines. This means $\overline{AC}$ is parallel to $\overline{BD}$. Work through a similar case with $\overline{AC}$ as the transversal that intersects $\overline{CD}$ and $\overline{AB}$.

Alternatively, you show that the opposite sides are parallel by selecting the segments and choosing **Measure | Slope.** In either case, ask, ***What other property can we add to our list?*** [Opposite sides are parallel.]

Properties of Rectangles
All angles are right angles (or, all angles are congruent).
Opposite sides are equal in length.
Diagonals are congruent and bisect each other.
Opposite sides are parallel.

11. Drag the rectangle so that it is a square, having four congruent sides. ***What shape does this look like?*** [Square] ***Is it still a rectangle? Explain.*** [Yes, it still has the properties of a rectangle.]

12. If time permits, discuss the Explore More. Have students review their construction methods and the properties they used in each. The last four pages of **Meet the Rectangle Present.gsp** show some methods that you may wish to share with students.

13. You may wish to have students respond individually in writing to this prompt. ***Compare a parallelogram and a rectangle. How are they similar? How are they different?*** [Both have opposite sides that are congruent and parallel. Both have diagonals that bisect each other. Both have adjacent angles that are supplementary. A rectangle has four congruent angles. A parallelogram has opposite angles that are congruent. A rectangle's diagonals are always congruent.]

EXTEND

Let students use their sketches of a rectangle to tell whether each statement is sometimes, always, or never true.

- A rectangle is a square. [Sometimes]
- A rectangle is a parallelogram. [Always]
- A rectangle is a quadrilateral. [Always]
- A rectangle is a polygon. [Always]
- A rectangle has four congruent sides. [Sometimes]
- A rectangle has only one pair of opposite, parallel sides. [Never]
- A rectangle has pairs of consecutive sides that are congruent and perpendicular. [Sometimes]

ANSWERS

14. Opposite sides of a rectangle are congruent.

19. Diagonals in a rectangle are congruent and bisect each other. Students might also notice that the diagonals in a rectangle divide the rectangle into two pairs of congruent isosceles triangles.

20. Answers will vary. Students may find one of the following methods.

Method: Construct $\overline{AB}$. Construct a line through point *A*, perpendicular to $\overline{AB}$. Construct $\overline{AC}$ on this line. Construct a circle with center *C* and radius $\overline{AC}$. Construct a circle with center *B* and radius $\overline{AB}$. Construct $\overline{BD}$ and $\overline{CD}$, where *D* is the intersection of the circles.
Properties: A rectangle has at least one right angle and the opposite sides are congruent.

Method: Construct $\overline{AB}$ and its midpoint, *C*. Construct circle *CA*. Construct $\overrightarrow{DC}$, where point *D* is on the circle. Construct point *E*, the intersection of the ray and the circle. Construct $\overline{AD}$, $\overline{DB}$, $\overline{BE}$, and $\overline{EA}$.
Properties: The diagonals of a rectangle are equal and bisect each other.

Meet the Rectangle

Name:

In this activity you'll construct a rectangle (a quadrilateral with four right angles). Then you'll discover some properties of rectangles.

CONSTRUCT

1. In a new sketch, construct a segment.

2. Label the endpoints *A* and *B*.

3. Now you'll construct lines perpendicular to $\overline{AB}$ through points *A* and *B*. Select point *A*, point *B*, and $\overline{AB}$ and choose **Construct | Perpendicular Lines.**

4. Construct a point on the line through point *A*.

5. Label it point *C*.

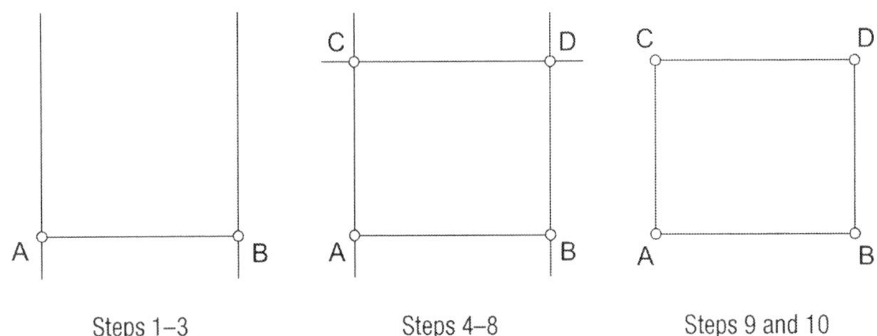

Steps 1–3 Steps 4–8 Steps 9 and 10

6. Construct a line through point *C* perpendicular to $\overleftrightarrow{AC}$.

7. Construct the point of intersection of this line and the line through point *B*.

8. Label it point *D*.

9. Select $\overleftrightarrow{AC}$, $\overleftrightarrow{CD}$, and $\overleftrightarrow{DB}$ and choose **Display | Hide Perpendicular Lines.**

10. Construct the missing segments $\overline{AC}$, $\overline{CD}$, and $\overline{DB}$.

11. Drag different vertices of your rectangle to make sure it's constructed properly.

EXPLORE

12. Measure the sides by selecting them and choosing **Measure | Length.**

13. Drag different parts of the rectangle and observe the measurements.

14. Make a conjecture about the sides of a rectangle.

CONSTRUCT

15. Construct the diagonals of the rectangle.

16. Construct the point of intersection of the diagonals.

17. Label it point *E*.

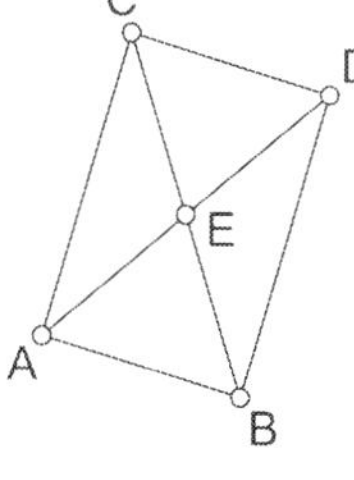

EXPLORE

18. Drag parts of the rectangle and observe the diagonals. Measure lengths that look as though they might be related.

    To measure a distance between two points, select the two points and then choose **Measure | Distance.**

19. Write at least two conjectures about the diagonals of a rectangle.

EXPLORE MORE

20. Find another way to construct a rectangle. Try methods that use the Construct menu, the Transform menu, or combinations of both. Consider how you might use diagonals. Write a brief description of the construction method along with the properties of rectangles that make that method work.

Meet the Rhombus: Properties of Rhombuses

INTRODUCE

Project the sketch for viewing by the class. Expect to spend about 10 minutes.

1. Open Sketchpad and enlarge the document window so it fills most of the screen.

2. Explain, ***Today you're going to use Sketchpad to construct a rhombus and investigate its properties. First I'll demonstrate how to construct a rhombus using its definition. As I construct the rhombus, think about how you would write the definition.*** As you demonstrate, make lines thick and labels large for visibility. Model the rhombus construction in worksheet steps 1–11 and the measurement of slope in worksheet step 12. Here are some tips.

   - In worksheet steps 1–3, demonstrate how to use the **Compass** tool to construct the circle. Explain that the radius point is a point on the circle that determines the length of the circle's radius. Then construct the radius using the **Segment** tool. Students can also select the center and radius point and choose **Construct | Segment.** Review the definition of *radius*: The length from the center of a circle to a point on the circle.

   - In worksheet steps 4–5, construct a second radius $\overline{AC}$. ***What do you know about the lengths of $\overline{AB}$ and $\overline{AC}$?*** [Because both are radii of the same circle, they are equal in length.]

   - In worksheet steps 6–8, explain that you will reflect $\overline{AB}$, $\overline{AC}$, and point A across $\overline{BC}$. ***$\overline{BC}$ will be the mirror line, or the line of reflection.*** Demonstrate how to use the **Arrow** tool to double-click it to mark it as a mirror. It will briefly animate to indicate that it has been marked. ***Predict what will happen when $\overline{AB}$, $\overline{AC}$, and point A are reflected across $\overline{BC}$.*** Listen to students' predictions, and then select point A, $\overline{AB}$, and $\overline{AC}$ and choose **Transform | Reflect.**

   - ***What do you know about the lengths of $\overline{A'C}$ and $\overline{A'B}$? Explain.*** [Because they are reflections of $\overline{AC}$ and $\overline{AB}$, and $\overline{AC}$ and $\overline{AB}$ are both radii of the same circle, the four segments are all the same length.]

   - In worksheet step 9, use the **Text** tool to label the reflection of point A as point A'. Explain that it is read "A prime." If students would prefer to label this as point D, tell them that they can double-click the label with the **Text** tool (or the **Arrow** tool), type a new label in the dialog box, and click OK.

- In worksheet step 10, demonstrate how to hide the circle and $\overline{BC}$ by selecting them and choosing **Display | Hide Path Objects.** ***How can we name this rhombus?*** [Rhombus *ABA′C* or *ACA′B*]
- In worksheet step 11, demonstrate how to drag the different parts of the rhombus to test the construction. Explain that this is an important step. If students make an error during the construction, remind them to use **Edit | Undo** as needed.
- Depending on the background of your students, you might want to review what *slope* means. Slope is the measure of the steepness of a line. It is the ratio of the rise (vertical change) to the run (horizontal change). ***What does it mean if two lines have the same slope?*** [They are parallel lines.]
- In worksheet step 12, model how to measure slope. A coordinate grid will appear. Show students how to hide the grid and axes by selecting them and choosing **Display | Hide.**

3. ***How does this construction help you define* rhombus?** Students may make the following response: *A rhombus is a shape with four sides all the same length.* Encourage students to add to the definition if it is not complete. ***How many sides does a rhombus have?*** [Four sides] ***What type of polygon is it?*** [Quadrilateral] ***What word describes a shape that has all equal sides?*** [Equilateral] Have students use these words in the definition, and then write it on chart paper: *A rhombus is an equilateral quadrilateral.*

4. If you want students to save their work, demonstrate choosing **File | Save As,** and let them know how to name and where to save their files.

DEVELOP

Expect students at computers to spend about 25 minutes.

5. Assign students to computers. Distribute the worksheet. Tell students to work through step 20 and do the Explore More if they have time. Encourage students to ask their neighbors for help if they are having difficulty with the construction.

6. Let pairs work at their own pace. As you circulate, here are some things to notice.

- In worksheet step 4, remind students that the circle will be highlighted when they have the **Segment** tool positioned on the circle. Have students test that the endpoint of the segment is on the circle by dragging the point with the **Arrow** tool.

- In worksheet steps 4 and 5, check students' understanding that all radii of a circle have equal length. ***Does where you construct point C on the circle have an impact on the length of $\overline{AC}$?***
- In worksheet step 8, if students have forgotten to mark $\overline{BC}$ as a mirror, Sketchpad will automatically mark a mirror. Have students choose **Edit | Undo** to go back a step and then correctly mark $\overline{BC}$ as a mirror.
- In worksheet step 11, make sure students complete the "drag test" to check that their constructions hold together properly.
- In worksheet step 12, remind students that they can hide the grid and axes. Encourage students to look for relationships among the slopes. ***What do you notice about the slopes of opposite sides?*** [They are equal.]
- In worksheet step 13, remind students to select the points that define an angle, with the second point being the vertex. Also, check that students don't measure the same angle twice. Encourage students to look for relationships among the angle measures. ***What do you notice about angles opposite each other? How about the consecutive angles, the angles next to each other?*** Students should observe that opposite angles have the same measure and adjacent angles are supplementary. Review the term *supplementary,* if needed. ***What are supplementary angles?*** [Angles whose measures sum to 180°.] ***Can you find any of these types of angles?***
- In worksheet step 14, as students drag parts of the rhombus, encourage them to make different types of rhombuses. ***Now make another rhombus that looks completely different. What changes as you drag parts of the rhombus? What stays the same?***
- In worksheet step 16, check that students understand that a diagonal is a line segment that connects two nonadjacent vertices. ***How will you construct the diagonals of the rhombus?*** [Construct $\overline{AA'}$ and $\overline{CB}$]
- In worksheet step 17, remind students that both diagonals will be highlighted when the point of intersection is chosen correctly.

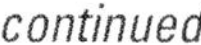

- In worksheet step 19, students will be measuring distances between points, although they could also construct new segments from the vertices to the point of intersection and then measure their lengths. Again, remind students to drag the parts of the rhombus to form many different rhombuses. Question students to help them see relationships as needed. ***What do you notice about the distances you measured?*** [They are equal.] ***What do you notice about the angle measures?*** [They are 90°.] ***What type of angles do the diagonals form when they intersect?*** [Right angles] ***What's a name for lines that intersect at 90-degree angles?*** [Perpendicular] ***What word describes dividing something into congruent parts?*** [Bisect] ***Can you use that word to describe something you see?*** [The diagonals bisect each other.]
- If students have time for the Explore More, they will use the properties of a rhombus to construct a rhombus, using a different method. Be sure students clearly describe their construction methods and explain what properties they used. Give students hints to start their thinking, if necessary. Have students drag vertices of their figures to make sure their constructions hold. Rhombuses that fall apart and can turn into other shapes are underconstrained. A construction that stays a rhombus but that can't take on all possible shapes of a rhombus is overconstrained.

7. If students will save their work, remind them where to save it now.

SUMMARIZE

Project the sketch. Expect to spend about 10 minutes.

8. Gather the class. Open **Meet the Rhombus Present.gsp** and use pages "Slopes and Angles" and "Diagonals" as needed. Students should have their worksheets with them. Begin the discussion by writing "Properties of Rhombuses" on chart paper. ***We started this activity by constructing a rhombus based on its definition. What property did we use to construct the rhombus?*** Write down the response on the chart paper.

Then review the conjectures students made about the sides, the angles, and the diagonals of rhombuses. Volunteers may come to the computer and drag the rhombus to show how the conjecture holds true for different rhombuses. Have students identify the properties that make a rhombus a parallelogram.

Properties of Rhombuses
All sides are congruent.
Opposite sides are parallel.
Opposite angles have equal measures.
Consecutive angles are supplementary.
Diagonals are perpendicular bisectors of each other.
Diagonals bisect the angles.

9. Drag the rhombus so that it is a square, having four congruent sides. ***What shape does this look like?*** [Square] ***Is it still a rhombus? Explain.*** [Yes, it still has the properties of a rhombus.]

10. If time permits, discuss the Explore More. Have students review their construction methods and the properties they used in each. The last four pages of **Meet the Rhombus Present.gsp** show some methods that you may wish to share with students.

11. You may wish to have students respond individually in writing to this prompt. ***Compare a rectangle and a rhombus. How are they similar? How are they different?*** [Both are parallelograms. A rectangle has four right angles. A rhombus has four congruent sides. The diagonals of a rhombus are perpendicular bisectors.]

EXTEND

Let students use their sketch of their rhombus to tell whether each statement is sometimes, always, or never true.

- A rhombus is a square. [Sometimes]
- A rhombus is a parallelogram. [Always]
- A rhombus is a quadrilateral. [Always]
- A rhombus is a polygon. [Always]
- A rhombus has four congruent sides. [Always]
- A rhombus is a rectangle. [Never]
- A rhombus has four congruent angles. [Sometimes]

- A rhombus has only one pair of opposite, parallel sides. [Never]
- A rhombus has pairs of consecutive sides that are congruent and perpendicular. [Sometimes]

ANSWERS

15. Opposite sides of a rhombus are parallel. Opposite angles of a rhombus are equal. Consecutive angles are supplementary.
20. Diagonals of a rhombus are perpendicular bisectors of each other. Diagonals of a rhombus bisect the angles.
21. Answers will vary. Students may find one of the following methods.

Method: Construct circles *AB* and *BA*. Construct a rhombus connecting the centers and the two points of intersection of the circles. (This is a special rhombus, composed of two equilateral triangles.)
Properties: A rhombus has four equal sides.

Method: Construct circles *AB* and *BA*. Construct circle *CA*, where point *C* is a point on circle *AB*. Construct point *D* at the point of intersection of circles *BA* and *CA*. *ABDC* is a rhombus.
Properties: A rhombus has four equal sides.

Method: Construct a circle *AB* and two radii. Construct a parallel line to each radius through the endpoint of the other.
Properties: A rhombus has equal consecutive sides and parallel opposite sides.

Method: Construct $\overline{AB}$ and its midpoint *C*. Construct a line through point *C*, perpendicular to $\overline{AB}$. Construct circle *CD*, where point *D* is on the perpendicular line. Construct point *E*, the other intersection of the circle with the line. *ADBE* is a rhombus.
Properties: The diagonals of a rhombus bisect each other.

Method: Construct a circle *AB* and point *C* on the circle. Bisect $\angle BAC$. Construct circle *BA* and point *D*, the intersection of this circle and the bisector. *ABDC* is a rhombus.
Properties: The sides of a rhombus are equal and the diagonals bisect the angles.

Meet the Rhombus

Name:

In this activity you'll construct a *rhombus* (a quadrilateral with four equal sides). Then you'll discover some properties of rhombuses.

CONSTRUCT

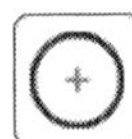

1. In a new sketch, construct a circle.

2. Label the center point *A* and the radius point *B*.

3. Construct radius $\overline{AB}$.
4. Construct another radius.

5. Label the new radius point *C*.

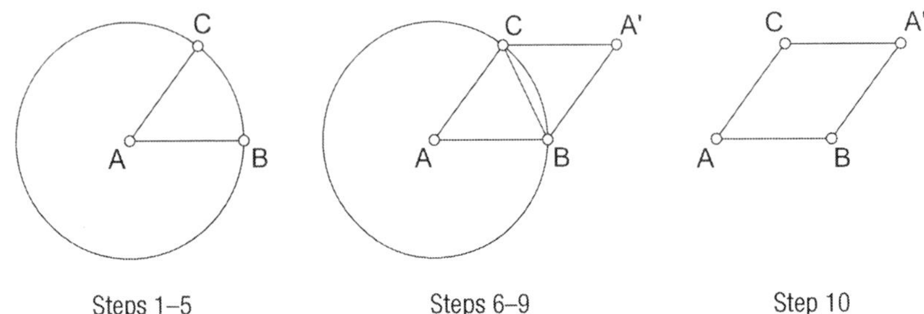

Steps 1–5 Steps 6–9 Step 10

6. Construct $\overline{BC}$.

Now you'll reflect point *A*, $\overline{AC}$, and $\overline{AB}$ across $\overline{BC}$.

7. Double-click $\overline{BC}$ to mark it as a mirror.
8. Then select point *A*, $\overline{AC}$, and $\overline{AB}$, and choose **Transform | Reflect.**

9. Label the reflection of point *A* as point *A′*.

10. Select the circle and $\overline{BC}$ and choose **Display | Hide Path Objects.**
11. Drag different vertices of your rhombus to make sure it's constructed properly.

EXPLORE

12. Select each of the sides and choose **Measure | Slope.**

13. Now measure the angles.

    To measure an angle, select three points with the vertex as the second point. Then choose **Measure | Angle.**

14. Drag the vertices and observe the measures.

15. Write at least three conjectures about the sides and the angles of a rhombus.

CONSTRUCT

16. Construct the diagonals of the rhombus.

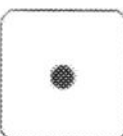

17. Construct the point of intersection of the diagonals.

18. Label it point *D*.

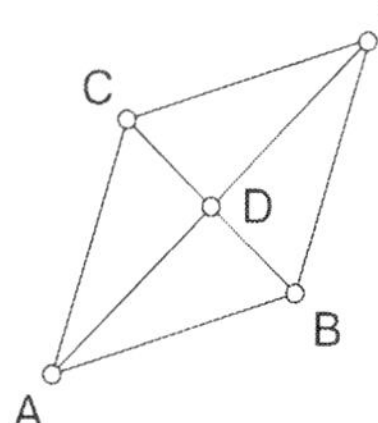

EXPLORE

19. Drag parts of the rhombus and observe how the diagonals are related to each other and to the angles in the rhombus. Measure lengths and angles that look as though they might be related.

    To measure a distance between two points, select the two points and then choose **Measure | Distance.**

20. Write at least two conjectures about the diagonals of a rhombus.

EXPLORE MORE

21. Find another way to construct a rhombus. Try methods that use the Construct menu, the Transform menu, or combinations of both. Consider how you might use diagonals. Write a brief description of the construction method along with the properties of rhombuses that make that method work.

Quadrilateral Pretenders: Classifying Quadrilaterals

ACTIVITY NOTES

INTRODUCE

Project the sketch for viewing by the class. Expect to spend about 10 minutes.

1. Open **Quad Pretenders Present.gsp** and go to page "Drag Test." Say, ***Some of these Sketchpad shapes have been constructed to always be squares, and some are just pretending to be.*** Drag vertices *A*, *B*, and *C* of the purple quadrilateral and see that it is no longer a square. ***What can you say about the shape?*** [It's always a parallelogram] Students may offer, among other ideas, that it is a rectangle, a parallelogram, or a quadrilateral. Suggest that they be as specific as possible and yet still include all possibilities. Help clarify that it's not a rectangle (although it may appear to be in some cases), but it's always a parallelogram.
2. As you drag other parts of the figure, ask, ***Does it stay a parallelogram no matter what I drag?*** [Yes] Students may need to revisit the definition of a parallelogram. ***The drag test showed that the parallelogram shape was pretending to be a square.*** Although the parallelogram initially looked like a square, it was not constructed to have the characteristics required in a square.
3. Review how a quadrilateral is named. ***One name for this parallelogram is parallelogram BCDA. What is another way to name this parallelogram?*** For example, students might say *ABCD, DABC,* or *CDAB.*
4. Refer to the blue quadrilateral. ***This second shape doesn't have its vertices named.*** Show students how to click a vertex with the **Text** tool to show its label. Once all the vertices are labeled, ask students to name the second quadrilateral by its vertices. For example, they might say *JKLI, LIJK,* and so on. ***What shape is IJKL?*** [A general quadrilateral. It has no more specific characteristics.]
5. ***What about shape EFGH? Is it always a square or is it a pretender?*** [It has a right angle.] Drag vertex *E*, and then vertex *F*. ***It still looks like a square. Let's try to drag the other vertices.*** When students see that it is no longer a square, ask them what kind of shape it is. Once they identify it as a trapezoid, ask, ***What else can you say about this trapezoid?*** [It has a right angle.] Encourage many responses. Students might say that it has two parallel sides, thus allowing you to remind them of the definition of a trapezoid. Bring out the fact that this particular trapezoid has a right angle.
6. ***What about shape MNOP? Is it always a square or is it a pretender?*** Drag each vertex in turn until students are satisfied that it will always

remain a square. ***So, shape MNOP is always a square, and the other three shapes were just pretending to be squares.***

7. As students begin work on the activity, let them know that identifying the pretenders is only part of the task; they must also describe as specifically as possible what each pretender really is. Encourage students to ask their classmates for help with Sketchpad, if needed.
8. If you want students to save their work, demonstrate choosing **File | Save As,** and let them know how to name and where to save their files.

DEVELOP

Expect students working at computers to spend about 25 minutes.

9. Assign students to computers and tell them where to locate **Quadrilateral Pretenders.gsp.** Distribute the worksheet. Let pairs work at their own pace. As you circulate, here are some things to notice.
   - For students who may be jumping to conclusions, ask, ***Have you tried to drag each vertex?***
   - If students drag a quadrilateral so that its edges cross, they no longer have a quadrilateral. ***For this activity you will need to limit your dragging to reshaping the quadrilateral without letting sides cross.***
   - Use questions to help students describe the pretenders completely.

     What stays the same as you drag this shape? What changes?

     How are these two opposite sides related?

     What is happening to this angle? To the other angles?
   - Quadrilaterals should be identified as specifically as possible. For example, on the "Trapezoid Pretenders" page, once students have identified trapezoids and pretenders, ask, ***Which quadrilaterals are trapezoids but also have more specific names?***
10. If students will save their work, remind them where to save it now. You might collect students' Explore More work on a flash drive or ask them to save their sketches where they can be displayed later.

SUMMARIZE

Expect to spend at least 10 minutes.

11. Gather the class. Students should have their worksheets with them. You might invite selected pairs to demonstrate their work on pages "Rectangles," "Parallelograms," and "Trapezoids." Ask demonstrators

to explain why they made particular choices. Alternatively, you might lead a class discussion around the solutions provided in **Quad Pretenders Present.gsp.**

12. If you have time, discuss the Explore More. Once students have results that pass the drag test, ask them to share these and the pretenders they made, and tell how they used the properties of the specific quadrilateral to construct the shape in Sketchpad. You might spend another day with students showing their construction work. These constructions could also extend to projects that are presented later.

13. ***What have we learned through this investigation?*** Help students articulate whatever they have learned. Include the objectives of the lesson. Students might offer suggestions like these, or you might want to bring these up.

    - The drag test reveals pretenders—shapes that may *look* like a particular shape, but are not constructed to have the required characteristics.
    - A general quadrilateral can pretend to be any of the more specific quadrilaterals.
    - The square fits within all definitions if your curriculum uses an inclusive definition of trapezoid (a quadrilateral with *at least* one pair of parallel sides). If your curriculum uses an exclusive definition of trapezoid (a quadrilateral with *exactly* one pair of parallel sides), then a square is a rectangle, a rhombus, a parallelogram, a kite, a right quadrilateral, and a quadrilateral.
    - Sometimes definitions of shapes are different (for example, trapezoid). It is important to understand the definitions, particularly when you are building a hierarchy of shapes.
    - A square could be called a right rhombus or an equilateral rectangle.
    - A rhombus is an equilateral parallelogram.
    - A rectangle is a right parallelogram.
    - A kite sometimes looks like a dart. ***What is the difference? Are their mathematical definitions different?*** Let students suggest definitions to distinguish the two shapes. Most likely they will involve an interior angle of more than 180° or use the term *concave.*

14. ***What other questions might you ask about relationships among shapes or definitions of shapes? You may or may not be able to answer them.*** Here are some ideas.

    Are there other kinds of quadrilaterals besides those we've discussed today?

    Is there a right kite? How many right angles would it have? [It could have exactly one right angle or all four right angles (making it a square).]

    Is there a special name for a trapezoid that has three sides the same length? (That is, one base is the same length as the sides.) [No.]

    Is it true that if shape A can pretend to be shape B (but is not actually shape B), then all B shapes are A shapes? [Yes. For example, a rectangle can pretend to be a square, and all squares are rectangles.]

    Can you classify shapes other than quadrilaterals, like triangles? [Yes. For example, equilateral, isosceles, and scalene triangles. Most other polygons are either regular (equilateral and equiangular) or irregular.]

EXTEND

Have students create a tree diagram or Venn diagram showing the relationships among quadrilaterals. For manageability, it's best not to include the right trapezoid or right quadrilateral in all diagrams. Including the kite is also problematic in a Venn diagram that uses inclusive definitions. Creating a visual display of the hierarchy of quadrilaterals could be a homework assignment, or you might create the diagram as a class. You might draw diagrams in Sketchpad like those presented in the last four pages of **Quad Pretenders Present.gsp.**

ANSWERS

1. "Parallelogram Pretenders": *JKLM:* kite; *NOPQ:* parallelogram; *CDEF:* rectangle; *RSTU:* quadrilateral; *VWXY:* trapezoid. *NOPQ* and *CDEF* are parallelograms. A rectangle is also a parallelogram.
2. "Rhombus Pretenders": *ABCD:* trapezoid; *RUST:* rhombus; *EFGH:* rectangle; *VWXY:* parallelogram; *NOPQ:* kite (or dart). *RUST* is a rhombus.
3. "Trapezoid Pretenders": *ABCD:* trapezoid; *JKLM:* quadrilateral; *NOPQ:* rectangle; *EFGH:* isosceles trapezoid. *NOPQ* and *EFGH* are isosceles trapezoids if you are using an inclusive definition of trapezoid.

4. "Square Pretenders": *ABCD:* square; *WXYZ:* rhombus; *KLMN:* parallelogram; *STUV:* rectangle; *CRAB:* quadrilateral; *FISH:* kite (or dart). *ABCD* is a square.

5. "Rectangles": All quadrilaterals can pretend to be rectangles, but only the square and rectangle are truly rectangles. See **Quad Pretenders Present.gsp.**

6. "Parallelograms": All quadrilaterals can pretend to be parallelograms, but the square, rectangle, rhombus, and parallelogram are truly parallelograms. See **Quad Pretenders Present.gsp.**

7. "Trapezoids": All quadrilaterals can pretend to be trapezoids. If you use the inclusive definition of trapezoid (having *at least* one pair of parallel sides), then the square, rectangle, rhombus, parallelogram, isosceles trapezoid, right trapezoid, and trapezoid are all truly trapezoids. If you use the exclusive definition of trapezoid (having *exactly* one pair of parallel sides), then only the last three of these quadrilaterals are trapezoids. See **Quad Pretenders Present.gsp.**

8. "Make Your Own": Constructions will vary. The drag test will indicate whether students have successfully created a specific type of quadrilateral or a pretender.

Quadrilateral Pretenders

Name:

If you see a shape that looks like a square, how do you know whether it's really a square? In Sketchpad, you can use the drag test. This test will show you those shapes that really are what they appear to be, and those that are only pretending.

EXPLORE

1. Open **Quadrilateral Pretenders.gsp.** Go to page "Parallelogram Pretenders." Of the five shapes that look like parallelograms, only two are constructed to always have opposite sides parallel. Drag the vertices to find them and match each shape to its most specific name.

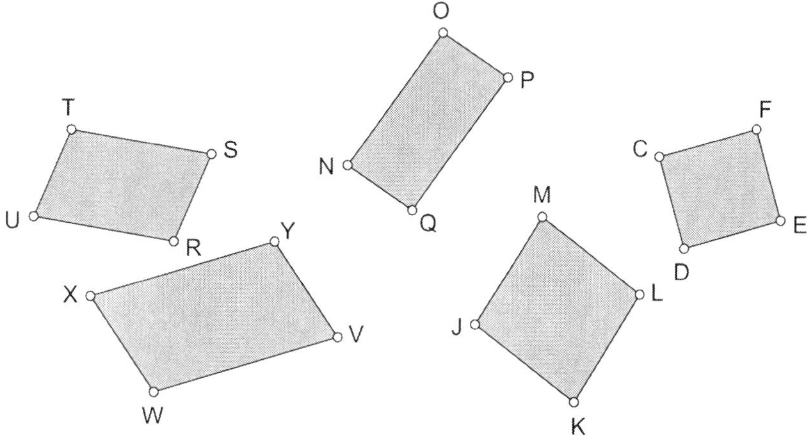

JKLM	parallelogram
NOPQ	trapezoid
CDEF	kite
RSTU	rectangle
VWXY	quadrilateral

Which two quadrilaterals are parallelograms?

What shape is a type of parallelogram but also has a more specific definition?

2. Go to page "Rhombus Pretenders." Of the five shapes that look like rhombuses, only one is constructed to always have four congruent sides. Find it, and identify each shape with its most specific name.

ABCD ______________ *RUST* ______________

EFGH ______________ *VWXY* ______________

NOPQ ______________

Quadrilateral Pretenders

continued

3. Go to page "Trapezoid Pretenders." Of the four shapes that look like isosceles trapezoids, only one is constructed to always be an isosceles trapezoid. Find it, and identify each shape with its most specific name.

   ABCD ____________ *JKLM* ____________

   NOPQ ____________ *EFGH* ____________

4. Go to page "Square Pretenders." Of the six shapes that look like squares, only one is constructed to always be a square. Match each shape to its most specific name.

ABCD	rhombus
WXYZ	kite
KLMN	quadrilateral
STUV	rectangle
CRAB	parallelogram
FISH	square

5. Go to page "Rectangles." Drag shapes that are always rectangles to the top of the page. Drag shapes that can pretend to be rectangles to the bottom of the page and make them look like rectangles.

6. Go to page "Parallelograms." Drag shapes that are always parallelograms to the top of the page. Drag shapes that can pretend to be parallelograms to the bottom of the page and make them look like parallelograms.

7. Go to page "Trapezoids." Drag shapes that are always trapezoids to the top of the page. Drag shapes that can pretend to be trapezoids to the bottom of the page and make them look like trapezoids.

EXPLORE MORE

8. Go to page "Make Your Own." Pick a type of quadrilateral and construct it. Then make a pretender of that shape. For example, you might make a square and a rhombus, or you might make a rectangle and a parallelogram.

   Shape ____________

   Pretender's shape ____________

All Things Being Equal: Constructing Regular Polygons

INTRODUCE

Project the sketch for viewing by the class. Expect to spend about 15 minutes.

1. Open Sketchpad and enlarge the document window so it fills most of the screen.
2. Explain, ***Today you're going to use Sketchpad to construct regular polygons using rotations. There are two construction methods you can use. I'll demonstrate one way, and then you'll try both methods on your own.***
3. Remind students that a *polygon* is a closed plane figure formed with three or more line segments as sides. Then ask, ***What is a regular polygon?*** Work with students to create a common definition and then write it on the board. Here is a sample definition: *A regular polygon is a polygon whose sides all have equal length and whose angles all have equal measure.* You might use this opportunity to remind students what *congruent* means.
4. Remind students that polygons are named by the number of sides. Review the names of common regular polygons: equilateral triangle (3), square (4), regular pentagon (5), regular hexagon (6), regular heptagon (7), regular octagon (8), regular nonagon (9), regular decagon (10), and regular dodecagon (12).
5. ***I'm going to model how to construct a regular pentagon by rotating a vertex point about the center point, using the measure of the central angle as the angle of rotation. What is a central angle?*** Review the definition with students: *A central angle is the angle formed by drawing lines from any two adjacent vertices to the center of the polygon.* Now model the regular pentagon construction in worksheet steps 1–9. If your students have some experience with Sketchpad, you might only model worksheet steps 1–5. As you demonstrate, make lines thick and labels large for visibility. Here are some tips.
   - In worksheet step 1, show students how to set up their preferences so points are automatically labeled. Explain that students can use the **Text** tool to edit labels, if needed, by double-clicking on the label of the object and entering a new label in the dialog box.
   - In worksheet step 2, construct point *A*. ***Point A will be the center point of the pentagon.*** Then construct point *B*. ***Point B will be one of the vertices of the pentagon. In this method we'll construct the other vertices of the pentagon by rotating point B about point A.***

- Mark point *A* as the center of rotation by double-clicking it with the **Arrow** tool. It will briefly animate, indicating that it has been marked. Alternatively, you can select point *A* and choose **Transform | Mark Center.**
- ***Because we want point B to rotate about point A, select point B with the* Arrow *tool and then choose* Transform | Rotate.** The Rotate dialog box will appear. ***What angle measure should we enter in the dialog box?*** Students may not have a response at this point. Ask questions to stimulate students' thinking. ***How many degrees are there around the center point A?*** [360°] Help students see that if you go completely around the center point *A*, you are going in a complete circle. ***How many vertices does a regular pentagon have?*** [Five] ***Because the lengths of the sides of a regular polygon are equal, the vertices are evenly spaced. How many degrees does a central angle have?*** To find the central angle, students can divide 360° by 5 to get 72°. Enter 72 for the angle measure and click **Rotate.**
- Explain that the new image is vertex *B′* and is read "*B* prime." Continue constructing the vertices of the regular pentagon.
- Demonstrate how to measure the side lengths and the interior angle measures. If needed, define *interior angle.* An interior angle is the angle formed by a pair of adjacent sides. Remind students that when measuring and naming angles, the vertex is the second point.
- Verify that the pentagon you've constructed is a regular polygon according to the definition. Drag the vertices to test that the construction holds.

6. If you want students to save their work, demonstrate choosing **File | Save As,** and let them know how to name and where to save their files.

DEVELOP

Expect students at computers to spend about 35 minutes.

7. Assign students to computers. Distribute the worksheet. Tell students to work through step 28 and do the Explore More if they have time. If you are going to assign different polygons to each student pair, let each pair know which regular polygons they should construct. Encourage students to ask their neighbors for help if they are having difficulty with the construction.

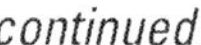

8. Let pairs work at their own pace. As you circulate, here are some things to notice.

   - In worksheet step 10, remind students that the drag test is important to test the construction. If the construction falls apart, have students go back one or more steps by choosing **Edit | Undo.** If the construction holds, students should drag the vertices to create different-size pentagons to check that the properties for a regular pentagon hold true for any regular pentagon: all sides have equal length and all interior angles have equal measure.
   - In worksheet step 14, angle BAB' is a central angle. Review the definition with students as needed. The measure of the central angle was used to construct the regular pentagon; this was the fixed angle of rotation.
   - In worksheet steps 15–22, students will construct a regular pentagon by using an interior angle measure as the fixed angle of rotation. Ask students to explain why they think 108° is used for the interior angle measure. Depending on the background of your students, they might use the formula for the sum of the interior angles of a polygon with n sides, $180(n - 2)$, and divide this sum by the number of sides to find the interior angle measure. In this case, the equation would be $180(5 - 2) \div 5 = 108$. Alternatively, they might notice that they can subtract the central angle measure from 180°.
   - As students are working on this construction method, remind them that, unlike in the previous construction, they need to mark a new center of rotation and select a new vertex to rotate each time.
   - In worksheet step 26, students will use the two construction methods to construct other regular polygons. Students may not get through all the constructions, depending on the amount of computer time they have. For the constructions they do complete successfully, students can make custom tools of them to use in later work. To make a custom tool, students must select the entire construction and then press and hold the **Custom** tool icon. From the Custom Tools menu, they choose **Create New Tool**. A dialog box will appear, in which students can type a name for their tool.
   - If students have time for the Explore More, they will construct a regular heptagon. Students will need to rotate by a marked angle rather than a fixed angle, because the angle measure is an infinite

repeating decimal (900° ÷ 7 for an interior angle and 360° ÷ 7 for a central angle). If students try to rotate by a fixed angle, they can enter only a decimal approximation in the Rotate dialog box. After seven rotations, the error of this approximation may be enough that the resulting polygon won't be perfectly regular. You might suggest that students use three decimal places. When using the Sketchpad Calculator to calculate the angle measure, students should choose **degrees** in the Units menu so that Sketchpad will recognize the calculation as an angle measure. Students can select this measurement and choose **Transform | Mark Angle** to mark it as the angle of rotation.

9. If students will save their work, remind them where to save it now.

SUMMARIZE

Project the sketch. Expect to spend about 10 minutes.

10. Gather the class. Students should have their worksheets with them. Begin the discussion by asking students to share which regular polygons they constructed using which method. ***How are the construction methods different?*** Students should explain that one method uses a central angle as the angle of rotation and the other method uses an interior angle as the angle of rotation. ***Why is it possible to use these angle measures to create a regular polygon?*** Bring out the idea that because the angle measures are equal in a regular polygon, you can use the angle measure as the angle of rotation to construct the other angles.

11. ***How can you find the central angle measure for any regular polygon with n sides?*** [Divide 360° by the number of sides.] ***How can you find the interior angle measure for any regular polygon with n sides?*** [Find the sum of the interior angles, then divide the sum by the number of sides: $180(n - 2) \div n$.] ***What is the relationship between the central angle and the interior angle in any regular polygon?*** [They are supplementary angles.]

12. ***Suppose you construct more regular polygons and keep increasing the number of sides each time. What figure does the regular polygon start to look like as the number of sides increases?*** [The figure approaches a circle.]

13. ***Is an interior angle in a regular pentagon larger or smaller than an interior angle in a regular octagon? Is the central angle in a regular pentagon larger or smaller than a central angle in a regular octagon?***

What general statement can you make about the sizes of the interior angle and the central angle as the number of sides in a regular polygon increases? You may wish to have students respond individually in writing to this prompt. Students should notice that as the number of sides increases, the interior angle increases and the central angle decreases.

14. If time permits, discuss the Explore More. Have students explain the difference between using a marked angle and a fixed angle of rotation. Then ask students to give the measures of the interior angle and of the central angle.

EXTEND

1. Ask students to construct a square and a regular octagon using this method.
   - Construct a circle using the **Compass** tool.
   - Construct a diameter.
   - Construct a diameter perpendicular to the first diameter.

   The points of intersection with the circle are the vertices of the square. Students can continue on to construct the regular octagon by finding the angle bisectors of the central angles. Encourage students to find other methods to construct regular polygons.
2. Have students explore the exterior angles and their relationship to the interior and central angles. Exterior and interior angles are supplementary. Exteriors and central angles are congruent. You might have students try to explain why this is true.

ANSWERS

11. The lengths of the sides are all equal; the measures of the angles are all equal.
12. Yes, it meets the definition for a regular pentagon. The side lengths are all equal and the measures of the interior angles are all equal.
13. One full revolution is 360°, so one-fifth of a revolution is 360° ÷ 5, which is 72°.
14. 72°; central angle

23. This construction method uses rotations to construct congruent interior angles.

24. The lengths of the sides are all equal.

25. Yes, it meets the definition for a regular pentagon. The side lengths are all equal and the measures of the interior angles are all equal.

26.

Polygon	Number of Sides	Central Angle Measure	Interior Angle Measure
Equilateral Triangle	3	120°	60°
Square	4	90°	90°
Regular Pentagon	5	72°	108°
Regular Hexagon	6	60°	120°
Regular Octagon	8	45°	135°
Regular Nonagon	9	40°	140°
Regular Decagon	10	36°	144°
Regular Dodecagon	12	30°	150°

27. $360° \div n$

28. $180(n - 2) \div n$

29. Central angle $= 51.\overline{428571}°$; interior angle $= 128.\overline{571428}°$

All Things Being Equal

Name:

A regular polygon is a polygon whose sides all have equal length and whose angles all have equal measure. In this activity you'll construct regular polygons using rotations.

CONSTRUCT

First you'll construct a regular pentagon by rotating vertex point *B* around center point *A*. You'll rotate by the measure of the central angles.

1. In a new sketch, choose **Edit | Preferences.** On the Text Panel, under **Show labels automatically,** check **For all new points.** Click **OK.**

2. Construct points *A* and *B*.

3. Double-click point *A* to mark it as the center of rotation.
4. Select point *B*, and choose **Transform | Rotate.** In the dialog box, set Angle to 72°, and then click **Rotate.**
5. Select the new image, point *B'*, and rotate it 72°.
6. Continue to rotate each new image 72° until you have five points.

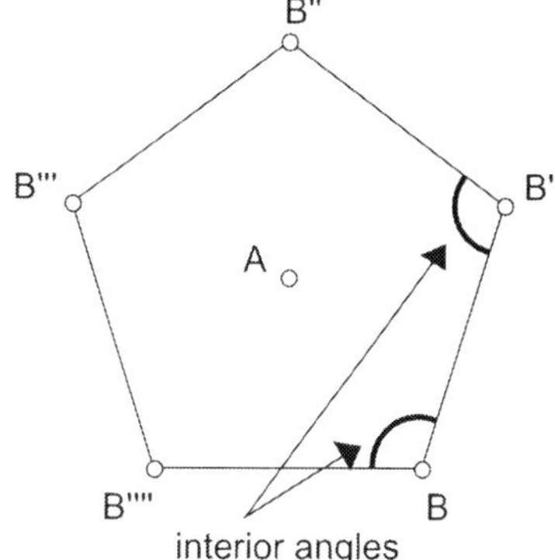

7. Connect all five points in order with segments.

8. Measure the sides by selecting them and choosing **Measure | Length.**
9. Now you'll measure the interior angles. Select three points of an interior angle with the vertex as the second point. Choose **Measure | Angle.**

EXPLORE

10. Drag the vertices to create different-size pentagons.

11. What do you observe about the lengths of the sides and the measures of the angles?

12. Does the definition for regular polygon hold true for your pentagon? Explain.

13. Why does repeatedly rotating a point by 72° around the center result in a regular pentagon?

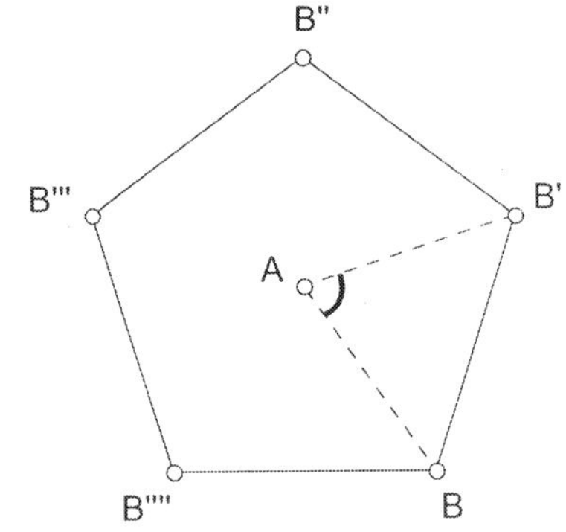

14. What is the measure of $\angle BAB'$? What is this angle called?

CONSTRUCT

Now you'll construct a regular pentagon by rotating vertex point *B* around another vertex point *A*. You'll rotate by the measure of the interior angles.

15. On a new page **(File | Document Options)**, construct points *A* and *B*.

16. Mark point *A* as the center of rotation.

17. Select point *B* and choose **Transform | Rotate.** In the dialog box, enter the value you found for the interior angles in step 9. Then click **Rotate.**

18. Mark the new image, point *B'*, as the center of rotation.

19. Select the previous center of rotation, point *A,* and rotate it by the same value as in step 17.

20. Continue to mark the new image as the center of rotation and to rotate the previous center of rotation until you have five points.

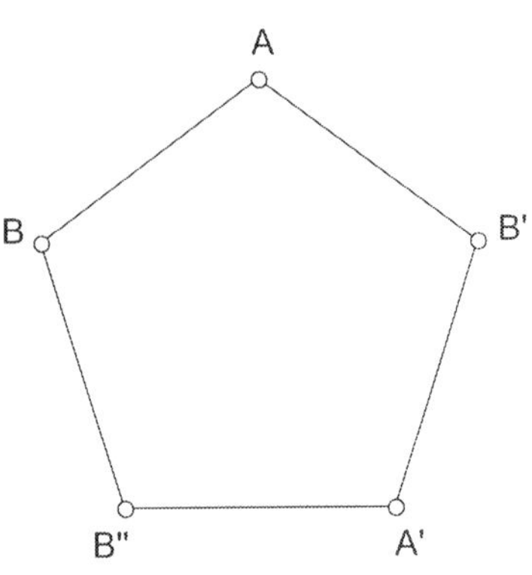

21. Connect all five points in order with segments.

22. Select the sides and choose **Measure | Length.**

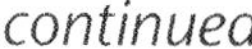

EXPLORE

23. Explain why you don't need to measure the interior angles to know that they are all congruent.

24. Drag the vertices to create different-size pentagons. What do you observe about the lengths of the sides?

25. Does the definition for *regular polygon* hold true for your pentagon? Explain.

CONSTRUCT

26. Experiment with using rotations to construct different regular polygons. Each time you construct a regular polygon that seems right, drag the vertices to make sure it holds together. Fill in the table with the central and interior angle measures for each regular polygon you construct.

Polygon	Number of Sides	Central Angle Measure	Interior Angle Measure
Equilateral Triangle			
Square			
Regular Pentagon			
Regular Hexagon			
Regular Octagon			
Regular Nonagon			
Regular Decagon			
Regular Dodecagon			

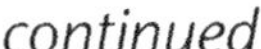

EXPLORE

27. Look at your table. How can you find the central angle measure of a regular polygon with *n* sides?

28. How can you find the interior angle measure of a regular polygon with *n* sides?

EXPLORE MORE

29. The regular heptagon (seven sides) doesn't appear in the table because the angle measures aren't "nice." What are they?

30. Construct the regular heptagon. Use the Sketchpad Calculator to calculate an expression for the desired angle, and then mark that measurement as an angle for rotation.

INTRODUCE

Project the sketch for viewing by the class. Expect to spend about 10 minutes.

1. Open **Polygon Pretenders Present.gsp** and go to page "Drag Test."
2. Say, ***Some of these Sketchpad shapes have been constructed to be regular polygons, and some are just pretending to be.*** Drag vertex *A* of the yellow pentagon and show that it is no longer regular. ***What stays the same as I drag? What changes? What can you say about the shape?*** [It's always an equilateral pentagon.] Students may offer that it's a pentagon. Suggest that they be as specific as possible. Help clarify that it's not always a regular pentagon (although it may pretend to be), but it's always an equilateral pentagon. Define *equilateral* (all sides congruent), if needed.
3. As you drag other parts of the yellow pentagon, ask, ***Does it stay an equilateral pentagon no matter what I drag?*** [Yes] ***The drag test showed that the equilateral-polygon shape was pretending to be regular.*** Although the yellow pentagon initially looked like a regular polygon, it was not constructed to always have all angles *congruent* (having equal measure). But it was constructed to always have all sides congruent.
4. Drag the vertices of the purple hexagon. ***Is the purple-hexagon shape always a regular polygon?*** [No] ***What can you say about this shape?*** [It's always an equiangular polygon.] The purple hexagon is constructed to have all angles congruent, but not all sides. Define *equiangular* (all angles congruent), if needed.
5. ***What about the orange quadrilateral—is this shape always a square or is it a pretender?*** Drag vertex *I, and* then vertex *F.* ***It still looks equilateral. Let's try to drag another vertex.*** When students see that it is no longer a square, ask them, ***What can you say about this shape?*** [It's always a right trapezoid.] Students should recognize that it is more specific than just a trapezoid.
6. Drag vertices of the green octagon to verify that it always remains a regular polygon. Drag vertices *Q* and *R* to resize the octagon. Define *regular* (all sides congruent and all angles congruent), if needed.
7. As students begin work on the activity, let them know that identifying the pretenders is only part of the task; they must also describe as specifically as possible what each pretender really is. Encourage students to ask their classmates for help with Sketchpad, if needed.

DEVELOP

Expect students at computers to spend about 25 minutes.

8. Assign students to computers and tell them where to locate **Polygon Pretenders.gsp.** Distribute the worksheet. Tell students to work through step 8 and do the Explore More if they have time.

9. Let pairs work at their own pace. As you circulate, here are some things to notice.

   - For students who may be jumping to conclusions, ask, ***Have you tried to drag each vertex?***

   - Use questions to help students describe the pretenders completely.

     What stays the same as you drag this vertex or edge? What changes?

     How are these two sides related?

     What is happening to this angle (side)? To the other angles (sides)?

   Encourage students to identify each shape as specifically as possible.

10. You might collect students' Explore More work on a flash drive or ask them to save their sketches where they can be displayed during the class discussion.

SUMMARIZE

Project the sketch. Expect to spend about 10 minutes.

11. Gather the class. Students should have their worksheets with them. You might invite selected pairs to demonstrate their work on worksheet steps 6–8. Ask demonstrators to explain why they made particular choices. Alternatively, you might lead a class discussion around the solutions provided in **Polygon Pretenders Present.gsp.**

12. If you have time, discuss the Explore More. Have students share their methods for constructing regular polygons of differing numbers of sides.

13. ***What have you learned through this investigation?*** Help students articulate whatever they have learned. Include the objectives of the lesson. Students might offer suggestions such as these, or you might want to bring these up.

    - The drag test reveals pretenders—shapes that may *look* like a particular shape, but are not constructed to have the required characteristics.

- An equilateral polygon has all sides congruent, but not necessarily all angles congruent.
- An equiangular polygon has all angles congruent, but not necessarily all sides congruent.
- A regular polygon is both equilateral and equiangular.

EXTEND

What other questions might you ask about relationships among shapes? Here are some ideas students might suggest.

Are the pretenders really pretending? Aren't they actually the shape they're pretending to be, even if only for that instant?

Are there other special kinds of polygons?

If shape A can pretend to be shape B, are all B-shapes a subset of A-shapes?

How can you construct an equilateral polygon that is not equiangular, or an equiangular polygon that is not equilateral?

ANSWERS

1. *PENTA* is the regular pentagon.
2. *DANCE* is the *SASAS* pentagon (three sides congruent and the included angles congruent).
3. *MNOPQ:* general pentagon; *SMILE:* equilateral pentagon; *VWXYZ:* equiangular pentagon; *PENTA:* regular pentagon; *DANCE:* SASAS pentagon
4. *LMNOPQ:* general hexagon; *ABCDEF:* regular hexagon; *FOREST:* equiangular hexagon; *TIGERS:* equilateral hexagon; *UVWXYZ:* SASAS hexagon
5. *ABCDEFGH:* octagon with opposite sides parallel; *JKLMNOPQ:* SASAS octagon; *STUVWXYZ:* general octagon; *MATHWORK:* equilateral octagon; *PLAYTIME:* equiangular octagon; *TENBLOGS:* regular octagon
6. The regular pentagon, regular hexagon, and regular octagon are on the top. The two equilateral polygons and two equiangular polygons are on the bottom, pretending to have all sides and angles congruent. See **Polygon Pretenders Present.gsp.**

7. The three regular polygons and two equilateral polygons are on the top. The two equiangular polygons are on the bottom, pretending to have all sides congruent. See **Polygon Pretenders Present.gsp.**

8. The three regular polygons and two equiangular polygons are on the top. The two equilateral polygons are on the bottom, pretending to have all angles congruent. See **Polygon Pretenders Present.gsp.**

9. Constructions will vary. The drag test will indicate whether students have successfully created a specific triangle and a pretender.

Polygon Pretenders

Name:

If you see a Sketchpad shape that looks like a regular polygon, how do you know whether it is constructed to always be regular? In this activity you'll use the drag test to find those shapes that are always what they appear to be and those that are only pretending.

EXPLORE

1. Open **Polygon Pretenders.gsp.** Go to page "Pentagon Pretenders." Of the five shapes that look like regular pentagons, only one is constructed to always have five congruent sides and five congruent angles. Drag the vertices to find it.

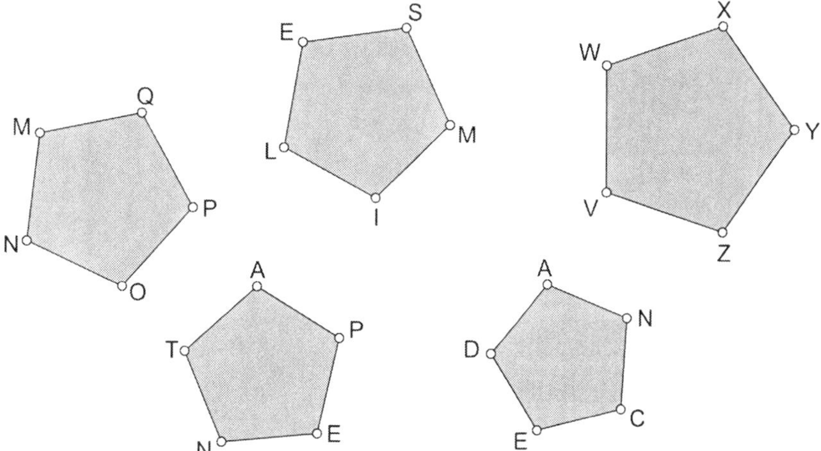

2. One of the pretenders has three adjacent sides that are congruent and two angles between those sides that are also congruent. We'll call shapes like these SASAS polygons. Drag the vertices to find it.

3. Identify each shape with its most specific name.

   General pentagon ________________

   Equiangular pentagon ________________

   Equilateral pentagon ________________

   Regular pentagon ________________

   SASAS pentagon ________________

4. Go to page "Hexagon Pretenders." Of the five shapes that look like equilateral hexagons, only two are constructed to always have all sides congruent. Find them and match each shape with its most specific name.

General hexagon	*ABCDEF*
Regular hexagon	*TIGERS*
Equilateral hexagon	*UVWXYZ*
Equiangular hexagon	*FOREST*
SASAS hexagon	*LMNOPQ*

5. Go to page "Octagon Pretenders." Of the six shapes that look like equilateral octagons, only two are constructed to always have all angles congruent. Find them and match each shape with its most specific name.

   General octagon ________________

   Equilateral octagon ________________

   Equiangular octagon ________________

   Regular octagon ________________

   SASAS octagon ________________

   Octagon with opposite sides parallel ________________

6. Go to page "Regular Polygons." Drag shapes that are always regular polygons to the top of the page. Drag polygons that can pretend to be regular polygons to the bottom of the page and make them look regular.

7. Go to page "Equilateral Polygons." Drag shapes that are always equilateral polygons to the top of the page. Drag polygons that can pretend to be equilateral polygons to the bottom of the page and make them look equilateral.

8. Go to page "Equiangular Polygons." Drag shapes that are always equiangular polygons to the top of the page. Drag polygons that can pretend to be equiangular polygons to the bottom of the page and make them look equiangular.

EXPLORE MORE

9. Go to page "Make Your Own." Use the Construct and Transform menus to construct any regular polygon. Then try to construct a regular polygon with more sides.

Area Formulas

Parallel Pairs: Parallelogram and Triangle Area

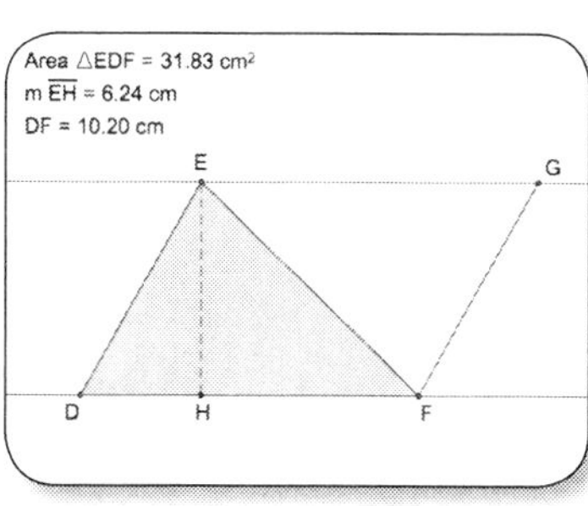

Students explore the relationship between the areas of parallelograms and triangles using a process called shearing. Students discover that shearing does not affect the area, but changing the lengths of the height and base does. Based on their observations, students write formulas for area of a parallelogram and area of a triangle.

Slanted Bases: Calculating Triangle Area

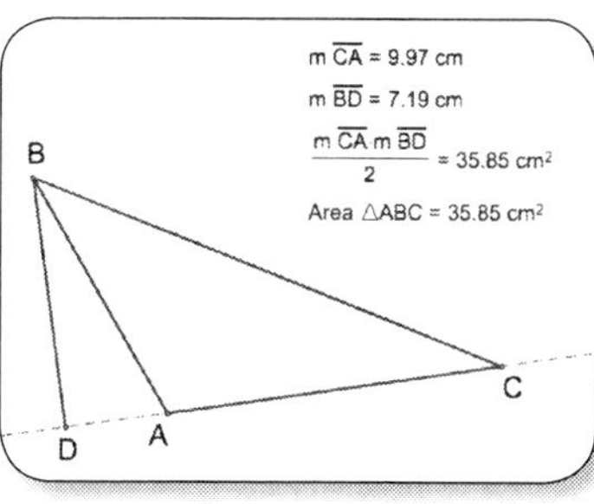

Students discover that they can calculate the area of a triangle using any side as a base. By manipulating a dynamic triangle, students explore altitudes, including those outside a triangle, and observe that the triangle's base and corresponding altitude do not have to be horizontal and vertical.

Sliding Bases: Parallelogram Area

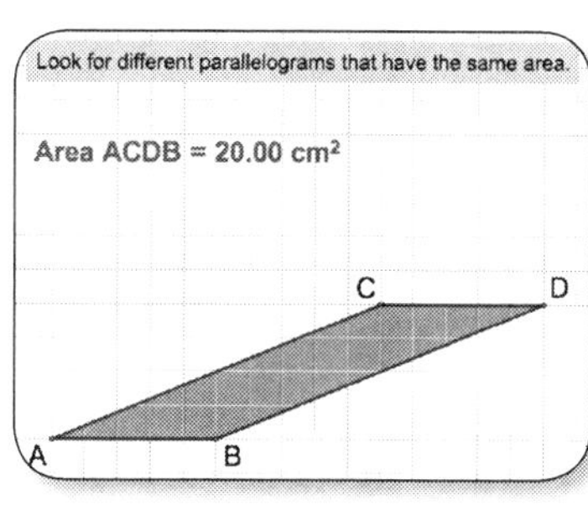

Students manipulate a dynamic parallelogram whose vertices have integer coordinates. They discover that by keeping the parallelogram's base and height measurements constant, they can create many parallelograms with the same area as a rectangle that has the same base and height. This discovery leads to the area formula for all parallelograms.

Shape Shearing: Parallelogram Area Formula

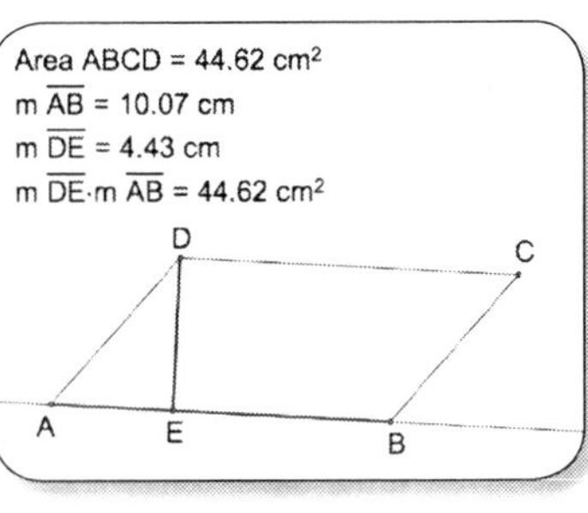

Students deepen their understanding of parallelogram area as they construct a segment representing the height, use the height to calculate area, and compare their calculation to the area measurement displayed by Sketchpad. Finally, students discover that dragging a vertex of the parallelogram without changing the base or height changes side lengths and the perimeter, but not the area.

One Parallel Pair: Trapezoid Area

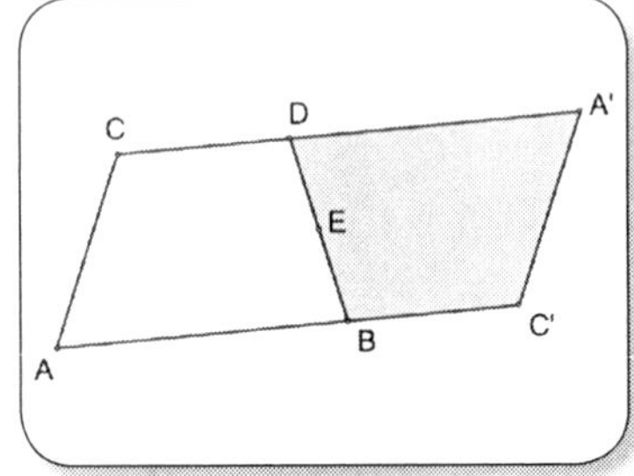

Students construct a trapezoid and rotate it 180° around a midpoint on its leg. Recognizing that a parallelogram is formed by the trapezoid and its rotated image, students use the area formula for a parallelogram to derive the area formula for a trapezoid.

Wavy Parallelogram: Circle Area

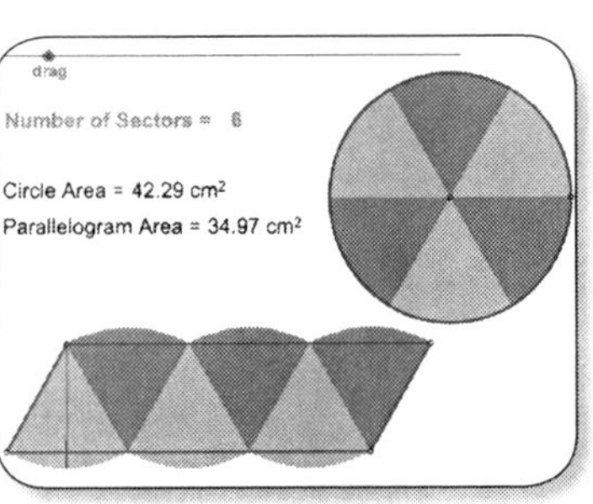

Students see an interactive model that relates circle area to parallelogram area by arranging circle sectors to create a "wavy parallelogram." Students use what they know about parallelogram area to derive the formula for circle area. This activity is led by the teacher; there is no student worksheet.

Smoothing the Sides: Regular Polygon and Circle Area

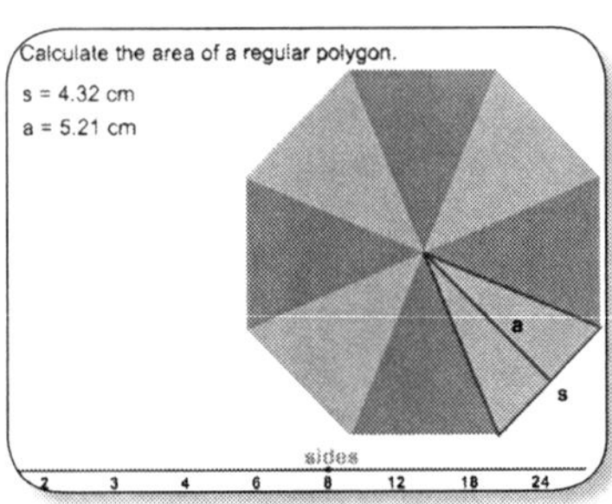

Students use a prepared sketch of a regular polygon that is divided into congruent triangles. Using the formula for area of a triangle, they write a formula for the area of any regular polygon. Then students observe how the area of the polygon approaches the area of the circumscribed circle as the number of sides increases, and use this to write a formula for the area of a circle.

Parallel Pairs: Parallelogram and Triangle Area

ACTIVITY NOTES

INTRODUCE

Project the sketch for viewing by the class. Expect to spend about 10 minutes.

1. Open Sketchpad and enlarge the document window so it fills most of the screen.
2. Explain, ***Today you're going to use Sketchpad to find the areas of parallelograms and triangles and to discover how they are related. You'll use a process called*** **shearing** ***to help you write the area formulas. First I'll demonstrate how to construct a parallelogram and then how to shear it.***
3. As you demonstrate, make lines thick and labels large for visibility. First model the parallelogram construction in worksheet steps 1–12. Then model how to shear the parallelogram in worksheet step 14. Here are some tips.

   - Start off by discussing the properties of parallelograms. ***How would you define a parallelogram?*** Listen for the following points.

     It's a polygon with four sides; it's a quadrilateral.

     The opposite sides are parallel.

     The opposite sides are congruent.

     Write a definition for *parallelogram* on chart paper. Here is an example: A parallelogram is a quadrilateral whose opposite sides are parallel.

   - In worksheet step 1, model how to set the Sketchpad Preferences so that points are automatically labeled. ***For this sketch, we'll have Sketchpad automatically label the points as we construct them. It will be easier to follow the construction.***

Instead of using the **Segment** tool to make side $\overline{FG}$, it may be easier for some students to select the points F and G and then choose **Construct | Segment.**

   - In worksheet steps 2–5, model how to make two parallel lines. ***We'll use these two lines to help us construct two sides of the parallelogram. What do we need to do next?*** [Make another pair of parallel lines to construct the other two sides of the parallelogram.]

   - Follow worksheet steps 6–12 to demonstrate how to make the other pair of parallel sides. ***How does this figure meet our definition of a parallelogram?*** [It has four sides and the opposite sides are parallel.] If your curriculum's definition of a parallelogram includes that the opposite sides are congruent, you can measure the lengths of sides or the distances between vertices to verify that this condition is met.

- ***What is the name of this figure?*** [Parallelogram *DEGF*] Remind students that parallelograms are named by the vertices listed in order.
- In worksheet step 14, demonstrate how to shear the parallelogram by dragging point *E*. ***What happens when I shear the parallelogram?*** [The shape of the parallelogram changes.] ***Is the figure still a parallelogram no matter where I drag point E? Explain.*** [Yes, because opposite sides are still parallel.] Tell students that they will explore shearing a parallelogram more on their own.

4. If you want students to save their work, demonstrate choosing **File | Save As,** and let them know how to name and where to save their files.

DEVELOP

Expect students at computers to spend about 35 minutes.

5. Assign students to computers. If you decide to have students skip the parallelogram construction, tell them where to locate **Parallel Pairs.gsp.** Distribute the worksheet. Tell students to work through step 28 and do the Explore More if they have time. Encourage students to ask their neighbors for help if they are having difficulty with the construction.
6. Let pairs work at their own pace. As you circulate, here are some things to notice.
   - In worksheet step 14, make sure students understand that in this construction, a base of the parallelogram is $\overline{DF}$ or $\overline{EG}$ and the height is the shortest distance between a base and its opposite side—a perpendicular segment. Ask students to observe what happens to the base and its height when they drag each time. ***What happens to the base when you drag point E?*** [Nothing, it stays the same.] ***What happens to the height?*** [Nothing, it stays the same.] ***Now drag point D toward point F. What happens to the base?*** [It gets shorter.] ***What happens to the height?*** [Nothing, it stays the same.] ***Now drag*** $\overline{EG}$ ***down. What happens to the base?*** [Nothing, it stays the same.] ***What happens to the height?*** [It gets shorter.] Let students finish exploring on their own.
   - In worksheets steps 16 and 17, students construct a height for the parallelogram. ***How do you know that*** $\overline{EH}$ ***is the height?*** [It is perpendicular to the base $\overline{DF}$.] *Note:* $\overline{EH}$ is also perpendicular to opposite side $\overline{EG}$; students may not see this initially.

- In worksheet step 18, be sure students select only the perpendicular line and not $\overline{EH}$. The perpendicular segment, or height, should still remain after hiding the perpendicular line. Show students how to choose **Edit | Undo** if they make an error.
- In worksheet step 20, students measure $\overline{DF}$, a base of the parallelogram. ***Would you get the same result if you measured the distance between points E and G? Explain.*** [Yes, because opposite sides are congruent] If necessary, have students measure the distance between vertices. ***What is another name for $\overline{EG}$?*** [It is also a base of the parallelogram.]
- In worksheet steps 21 and 22, listen to students as they derive the formula. ***Do you see a relationship between the base, the height, and the area? Go back and try dragging your parallelogram as described in worksheet step 14. What happens?***
- In worksheet step 25, ask students to define parts of $\triangle DEF$. ***What does $\overline{EF}$ represent?*** [A side of the triangle] ***What does $\overline{DF}$ represent?*** [A side and a base of the triangle] ***What does $\overline{EH}$ represent?*** [The height of the triangle] ***How do you know?*** [It is perpendicular to the base $\overline{DF}$.] Check that students understand that the triangle and the parallelogram share the same base and height.
- In worksheet step 27, question students as they drag point *E*. ***What happens to the base?*** [It stays the same.] ***To the height?*** [It stays the same.]
- If students have time for the Explore More, they will make an action button to animate point *E* along its line. The animation will shear the parallelogram.

7. If students will save their work, remind them where to save it now.

SUMMARIZE

Project the sketch. Expect to spend about 15 minutes.

8. Gather the class. Students should have their worksheets with them. Open **Parallel Pairs Present.gsp** and go to page "Parallelogram." Begin the discussion by having students identify the height and a base of the parallelogram. Write definitions for *height* and *base* on chart paper. Here are sample definitions: The height is the perpendicular distance from a base to the opposite side. The base is a side of the parallelogram. For your understanding, height is also referred to as "altitude."

9. Check students' understanding of these terms before proceeding. ***Is the base always the bottom of the parallelogram?*** [No, the base can be any side.] ***What is the height if $\overline{DE}$ is a base?*** [The perpendicular line segment from $\overline{DE}$ to the opposite side] ***Can the height lie outside of the parallelogram?*** [Yes] Demonstrate this concept by dragging points *D* or *F* to show the height outside of the figure.

10. Review worksheet step 15 with students. ***Did shearing the parallelogram change its area?*** [No] ***Why not?*** [The base and the height measurements stayed the same.] ***What measurements affect the area of the parallelogram?*** [The base and the height] ***Is this true for any size parallelogram?*** [Yes] Model this on the sketch.

11. Ask volunteers to share their formulas for area of a parallelogram. ***How did you arrive at your formula?*** Students' answers will vary. Here are some possible replies.

    The area changed when we changed the height and the base, so we looked for ways that the height and the base could equal the area.

    We noticed that the area is always greater than either the height or the base, so we tried adding the two. When that didn't work, we multiplied them.

    We noticed that you could make it into a rectangle, so we multiplied the base and the height.

    We changed the height and the base to whole number units so we could see the relationship more easily. When the height was 3 cm and the base was 4 cm, the area was 12 cm^2. We knew that $3 \times 4 = 12$, so we thought the area might be height times base. We verified our guess by trying other measurements and it worked!

    Use the Sketchpad Calculator to confirm that the product of the height and base is equal to the area. Write the formula on chart paper: $A = bh$.

12. Go to page "Triangle" and have students define the height and the base of $\triangle DEF$. ***What is the height of the triangle?*** [$\overline{EH}$] ***What is the base?*** [$\overline{DF}$] ***How do our definitions for height and base of a parallelogram compare to those for a triangle?*** [In both cases, the base is a side of the figure. The height of both figures is the perpendicular segment from a base; however, in a parallelogram, it is to the opposite side, and in a triangle, it is to the opposite vertex.] Write definitions for *height* and *base* of a triangle on chart paper: The height is the

perpendicular distance from a base to the opposite vertex (or the parallel line containing the opposite vertex). The base is a side of the triangle. Drag point *D* toward point *F* to model how the height is sometimes drawn outside the triangle.

13. Discuss worksheet steps 27 and 28. Drag point *E* in the example sketch. ***What happens to the area of the triangle when I drag point E?*** [The area stays the same.] ***Which measurements affect the area of the triangle?*** Drag points *D* or *F* or $\overline{DF}$. [The base and the height] ***How does this compare to what happens with a parallelogram?*** [The height and the base measurements affect the area of a parallelogram too.] ***What measurements do you need to find the area of a triangle?*** [The height and the base] ***How do you think the area of a triangle is related to the area of a parallelogram?*** Here are some possible responses.

    You use height and base measurements to find the areas of both figures.

    The triangle looks about half as big as the parallelogram, so take half of the parallelogram's area to find the triangle's area.

    If you divide the area of the parallelogram by 2, you get the area of the triangle.

    The area of the triangle is one-half the area of the parallelogram.

    Twice the area of the triangle equals the area of the parallelogram.

    Using what you observed, what is the formula for the area of a triangle? Write the area on chart paper: $A = \frac{bh}{2}$, or $\frac{1}{2}bh$.

14. You can verify this by showing the area of $\triangle EFG$ is the same as the area of $\triangle DEF$ and the sum of their areas is equal to the area of the parallelogram. First select points *E*, *F*, and *G* and choose **Construct | Triangle Interior.** Select $\triangle EFG$ and choose **Measure | Area.** Choose **Number | Calculate** and add the two triangle measurements. The sum will equal the area of the parallelogram.

15. You can also show that the triangle is half the area of the parallelogram by going to page "Double Triangle." Press *Double Triangle* to activate the animation.

16. ***How can you find the area of a triangle if you know the area of a parallelogram with the same base and height?*** You may wish to have students respond individually in writing to this prompt.

17. If time permits, discuss the Explore More. Students should note that the area remains constant because the height and the base do not change.

ANSWERS

15. The area of the parallelogram remains constant under shearing because the height and base of the parallelogram stay constant, and the area depends on only those two measures. Changing the base or the height of the parallelogram changes its area.
21. The area of the parallelogram equals $(m\overline{DF})(m\overline{EH})$.
22. $A = bh$
28. The area of the triangle is exactly half the area of the parallelogram.

$$A = \frac{bh}{2}$$

29. During the animation, the area of the parallelogram should not change because neither the height nor the base changes. Area remains constant under shearing for any figure because every cross section parallel to the shear line remains a constant length.

Parallel Pairs

For GSP5 Name:

In this activity you'll discover a relationship between the areas of parallelograms and triangles by investigating a process called *shearing*. This will help you write area formulas for parallelograms and triangles.

CONSTRUCT

1. In a new sketch, choose **Edit | Preferences | Text** and check **Show labels automatically: For all new points.** Click **OK.**

2. Construct a horizontal line, $\overleftrightarrow{AB}$.

3. Construct point *C* above $\overleftrightarrow{AB}$.

Steps 2–4

4. Now you'll construct a line parallel to $\overleftrightarrow{AB}$ through point *C*.

   Select point *C* and $\overleftrightarrow{AB}$, and then choose **Construct | Parallel Line.**

5. Hide points *A, B,* and *C* by selecting the points and choosing **Display | Hide Points.**

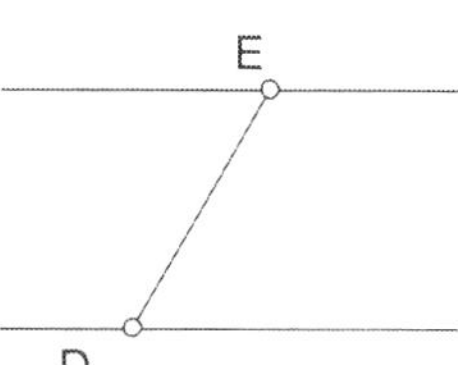

Steps 5 and 6

6. Construct $\overline{DE}$ from the bottom line to the top line.

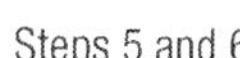

7. Construct point *F* on the bottom line.

8. Construct a line through point *F* parallel to $\overline{DE}$.

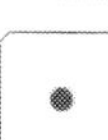

9. Construct point *G* where this line intersects the top line.

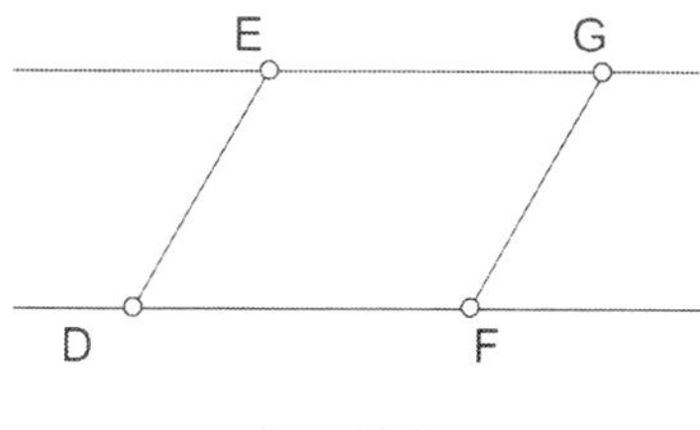

Steps 7–9

10. Construct interior *DEGF* by selecting the vertices in order and choosing **Construct | Quadrilateral Interior.**

11. Hide $\overleftrightarrow{FG}$.

12. Construct $\overline{FG}$.

EXPLORE

13. Measure the area of parallelogram *DEGF* by selecting the interior and choosing **Measure | Area.**

14. Observe the area measurement as you drag in each of these ways.
    - Drag point *E* to shear the parallelogram.
    - Drag point *D* or point *F* to change the base of the parallelogram.
    - Drag either $\overleftrightarrow{EG}$ or $\overleftrightarrow{DF}$ up or down to change the height.

15. Which of these actions change the area and which don't? Explain why you think this is so.

A height of a parallelogram is the shortest distance between two opposite bases. Now you'll construct an *altitude*—a segment that is perpendicular to the bases and whose length is a height of the parallelogram.

CONSTRUCT

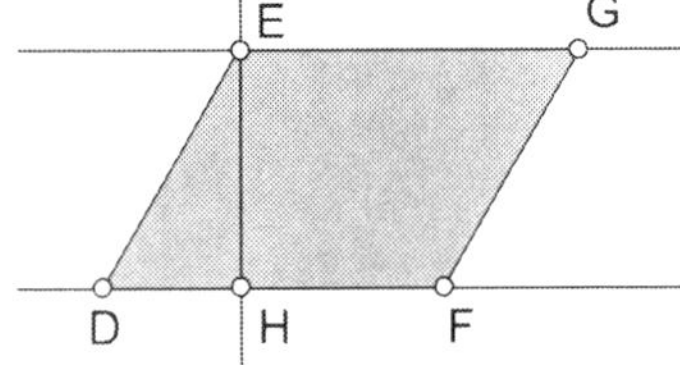

16. Select point *E* and $\overleftrightarrow{DF}$, and then choose **Construct | Perpendicular Line.**

17. Construct point *H* at the intersection of $\overleftrightarrow{DF}$ and the perpendicular line.

18. Hide the perpendicular line.

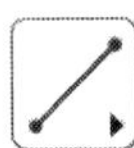

19. Construct base $\overline{DF}$ and altitude $\overline{EH}$.

EXPLORE

20. Select $\overline{DF}$ and $\overline{EH}$ and choose **Measure | Length.**

21. You will use these measurements to calculate an expression equal to the area of the parallelogram.
    Choose **Number | Calculate** to open the Sketchpad Calculator.
    Click once on a measurement to enter it into a calculation.
    Write down your expression.

22. Write a formula for the area of the parallelogram using *A* for area, *b* for base, and *h* for height.

Next, you'll investigate how the area of the parallelogram is related to the area of a triangle.

CONSTRUCT

23. Hide the interior of the parallelogram.

24. Construct diagonal $\overline{EF}$.

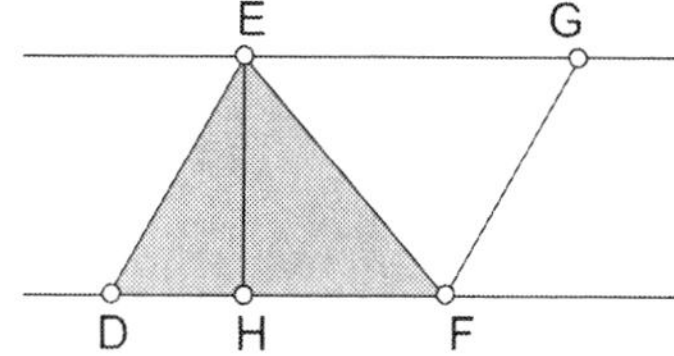

25. Construct triangle interior *DEF*.

EXPLORE

26. Measure the area of triangle *DEF*.

27. Drag point *E* and observe the area measurements.

28. How is the triangle's area related to the parallelogram's area? Write a formula for the area of the triangle using *A* for area, *b* for base, and *h* for height.

EXPLORE MORE

29. Make an action button to animate point *E* along its line. Explain what this animation demonstrates about shearing.

    Select point *E* and choose **Edit | Action Buttons | Animation.**

Slanted Bases: Calculating Triangle Area

ACTIVITY NOTES

INTRODUCE

Project the sketch for viewing by the class. Expect to spend about 10 minutes.

1. Introduce the activity. ***In this activity you'll explore the base, height, and area of a triangle that you'll construct.*** Review what students have learned about finding triangle area, including the formula $A = \frac{b \cdot h}{2}$. You might write the formula on the board. Don't bring up the idea that a triangle has three bases and three heights, but do entertain this point if students raise it, and let students know they'll be exploring this idea more using Sketchpad.

2. ***I'll model constructing a triangle in Sketchpad.*** Open page "Start" of **Slanted Bases Present.gsp** and enlarge the window so it fills most of the screen. As you demonstrate, make lines thick and labels large for visibility.
   - Use the **Segment** tool to draw the first side. Make the lines thick (**Display | Line Style**) for visibility.
   - Draw the other sides. Show students how to be sure that a vertex is highlighted when they click to connect another segment to it.
   - Drag a vertex to show that the triangle passes the drag test by remaining a triangle no matter how it is dragged.

3. ***You will be measuring the area of the triangle.***
   - If your students have not used the Calculator, model measuring the base of your triangle by selecting it and choosing **Measure | Length.** Then choose **Number | Calculate,** click on the measurement to insert it into the Calculator expression, and show students how to use the * on the Calculator to multiply.
   - In worksheet step 15, students check their calculations by measuring the area. Demonstrate how to construct the triangle interior by selecting the vertices, choosing **Construct | Triangle Interior,** and then choosing **Measure | Area.**

4. If students are to save their work, demonstrate how and where to save it.

DEVELOP

Expect students at computers to spend about 35 minutes.

5. Assign students to computers and distribute the worksheet. Tell students to work through all the steps of the worksheet and do the Explore More if they have time. Encourage students to ask a neighbor for help if they have questions about using Sketchpad.

If the labels appear automatically when you construct the segment, choose **Edit | Preferences.** In the Preferences dialog box, click the Text tab. For **Show labels automatically,** uncheck **For all new points.** Check **Apply to: New Sketches** and click **OK.** Then start again with a new sketch.

6. Let pairs work at their own pace. As you circulate, be alert to these issues.
   - In worksheet step 2, points may label automatically, depending on the setting in your software. Choosing **Edit | Preferences** will allow you to select to have Sketchpad show the labels of new points when they are created. Alternatively, students can use the **Text** tool to label points themselves.
   - In worksheet step 5, if you notice students are having trouble answering the question, ask, ***What do you notice about angles A and C when the perpendicular line falls outside the triangle?*** [One of these angles is obtuse.] Don't provide the answer. Students can move on if they don't see this right now.
   - In worksheet step 8, remind students to put their triangles to the drag test by dragging each of the vertices to make sure the line always intersects points *A* and *C*.
   - In worksheet step 11, if students have difficulty defining altitude, help them begin by providing the prompt, ***An altitude in a triangle is a line segment from . . . .*** Don't expect perfect definitions at this point. Students will have an opportunity to refine them in the whole-class discussion.
   - Worksheet step 17 illustrates an important objective of the activity, understanding that the base and altitude of a triangle need not be horizontal and vertical. Be sure students convince themselves of this fact by dragging their triangles sufficiently.

SUMMARIZE

Project the sketch for viewing by the class. Expect to spend about 15 minutes.

7. Students should have their worksheets. Open **Slanted Bases Present.gsp.** As you facilitate a class discussion of students' findings, invite volunteers to the computer to demonstrate.
8. Refer students to worksheet step 11. ***What are your ideas for a definition of* altitude?** [An altitude is a segment from the vertex of a triangle perpendicular to a line containing the base opposite that vertex.] Student definitions are apt to be less formal and may not include vocabulary such as *vertex* and *opposite*. However, make sure students understand the following two ideas.

- The altitude is perpendicular to the (line containing the) base.
- The altitude may not intersect the base, but instead may intersect the line extending the base.

A complete student definition will encompass answers to some of the previous worksheet questions.

9. ***What did you notice about the altitude's location and the type of angles (acute, obtuse, or right) at the triangle's base?*** Students should agree on the following points.
   - The altitude lies inside the triangle when both base angles are acute.
   - The altitude lies outside the triangle when a base angle is obtuse.
   - The altitude lies on a side of the triangle when a base angle is right.
10. ***Please share how you calculated the area of the triangle (step 13), and explain why that calculation worked.*** Students should have drawn on their previous understanding of the triangle area formula, $A = \frac{base \cdot height}{2}$, or $A = \frac{1}{2}base \cdot altitude$. They should have entered into the Calculator $\frac{1}{2}m\overline{AC} \cdot m\overline{BD}$ or $\frac{m\overline{AC} \cdot m\overline{BD}}{2}$.

You may also ask whether a triangle needs a "bottom" segment.

11. ***Does the base of a triangle have to be the bottom segment? What did you do today to answer this question?*** The students' discussion should include two observations.
    - Students moved the triangle's base and altitude to different positions.
    - Students constructed three altitudes, understanding that any side could serve as a base of the triangle.
12. Invite students who have completed the Explore More to share their additional observations about the three altitudes in a triangle.

ANSWERS

1. Most students will show the vertical altitude.
5. The perpendicular line doesn't intersect base $\overline{AC}$ when either $\angle A$ or $\angle C$ is obtuse or measures more than 90°.
6. The perpendicular line passes directly through point A when $\angle A$ is a right angle and directly through point C when $\angle C$ is a right angle.

11. An *altitude* in a triangle is a segment from the vertex of the triangle to a line containing the base opposite that vertex; the segment is perpendicular to that line.

15. Answers will very.

16. $A = \frac{base \cdot altitude}{2}$, or $A = \frac{1}{2} base \cdot altitude$.

17. The base is vertical.

22. The three calculations are equal. The area of a triangle can be calculated using any side as its base.

23. Students should identify all three sides of the triangle as bases and draw a corresponding altitude for each base.

24. Answers will vary depending on students' understanding at the beginning of the activity. Students should have learned that any side of a triangle can be a base, that for each base there is a corresponding altitude, and that an altitude can be outside the triangle.

25. In an acute triangle, the altitudes all lie inside the triangle. In an obtuse triangle, two altitudes lie outside the triangle. In a right triangle, two altitudes lie on the two sides forming the right angle, and the third lies inside. It is not possible for exactly two altitudes to lie inside the triangle; when one altitude is outside, another one is as well.

26. The lines through the three altitudes always intersect in a single point. This point, called the *orthocenter*, lies outside an obtuse triangle, inside an acute triangle, and at the vertex of the right angle in a right triangle.

Slanted Bases

Name:

In this activity you will explore the bases, heights, and areas of triangles.

1. Using a pencil, draw any heights and mark any bases on this triangle.

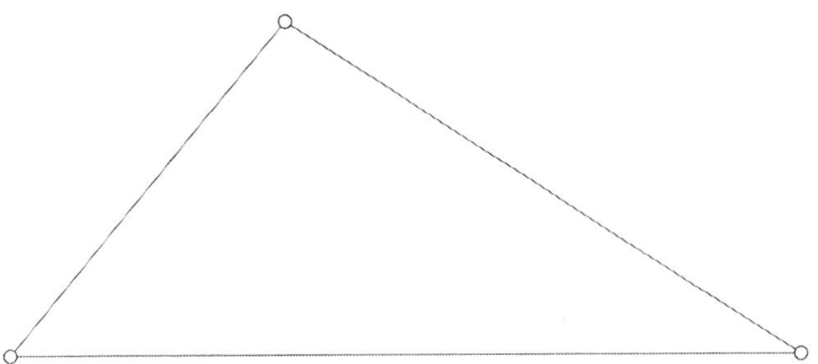

CONSTRUCT

2. Construct a triangle.

   Label the vertices *A, B,* and *C.*

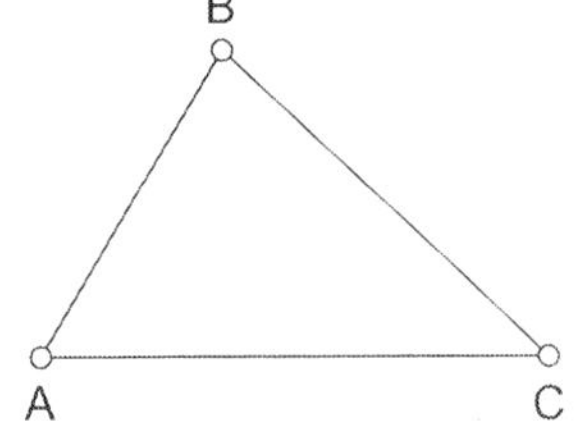

3. To construct an altitude from point *B* to base $\overline{AC}$, select *B* and $\overline{AC}$ and choose **Construct | Perpendicular Line.**

   The height of the triangle from point *B* to base $\overline{AC}$ is measured along this perpendicular line.

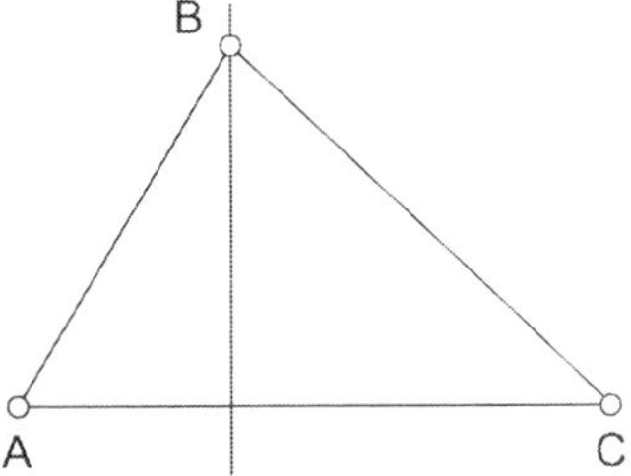

4. Drag point *B*. Note that the perpendicular line sometimes intersects base $\overline{AC}$ and sometimes doesn't.

5. What do you notice about the triangle when the perpendicular line doesn't intersect the base?

6. Describe the triangle when the perpendicular line passes directly through point *A* or point *C*.

Slanted Bases

continued

7. Drag point *B* so that the line does not intersect the base.

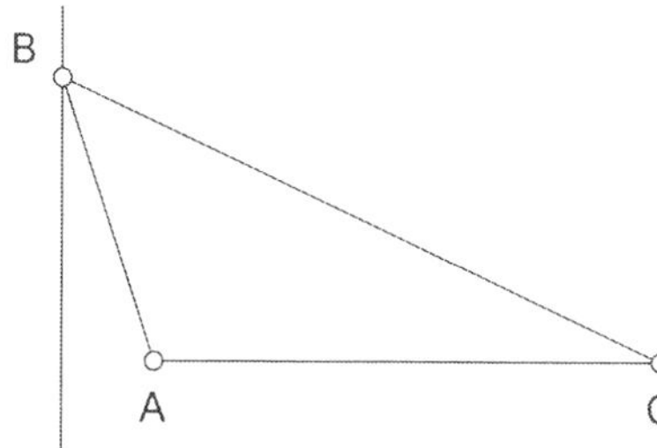

8. The perpendicular line and the base no longer intersect. To make them intersect again, you need to extend the base.

   Construct a line through points *A* and *C*.

   Drag the vertices to make sure the base and the perpendicular line always intersect.

9. Select the perpendicular line and the extended base $\overleftrightarrow{AC}$ and choose **Construct | Intersection.**

   Label the point of intersection *D*.

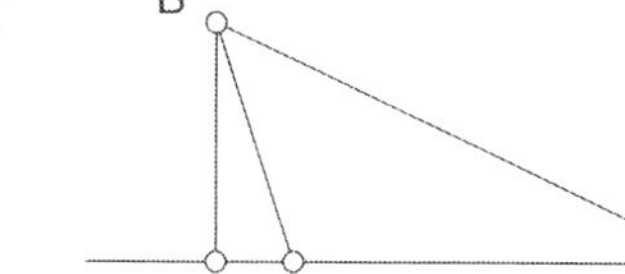

10. Hide the perpendicular line by selecting it and choosing **Display | Hide Perpendicular Line.**

   Construct a line segment between points *B* and *D*.

11. Segment *BD* is an *altitude* of the triangle.

   Use what you know about constructing $\overline{BD}$ to define *altitude*.

12. Measure the length of altitude $\overline{BD}$ by selecting it and choosing **Measure | Length.** Also measure the length of base $\overline{AC}$.

13. Now you will use the triangle area formula to find the area of this triangle. Choose **Number | Calculate.**

   Click on measurements in your sketch to insert them into the Calculator. (You may need to drag the Calculator to the side.) Then click **OK.**

14. Hide the measurements (not the calculation) by selecting them and choosing **Display | Hide Measurements.**

15. Now you will check your calculation.

    Select points *A, B,* and *C* and choose **Construct | Triangle Interior.**

    With the interior still selected, choose **Measure | Area.**

    Does the area measurement match your calculation? ____________

    If not, double-click your calculation. Change it so that it is correct. Drag vertices to confirm the two area measurements are the same.

16. Write the formula that tells how you calculated the area. Use *base* for the length of base $\overline{AC}$ and *altitude* for the length of $\overline{BD}$.

    Area = ____________

17. Drag the vertices and edges of the triangle.

    Sketch a triangle with its altitude horizontal.

    When the altitude is horizontal, what can you say about the base?

    Sketch a triangle that has its base on the top with the triangle pointing down.

Look at the questions in step 17 another way.

18. Use $\overline{AB}$ as the base and construct an altitude to it from vertex *C*. Start by extending the base, and use the same process you did before.

19. Measure altitude $\overline{CE}$. Measure base $\overline{AB}$.

    Use these lengths to calculate the area of the triangle.

    Hide the altitude and base measurements, but don't hide the calculations.

20. Now use $\overline{CB}$ as the base and construct an altitude $\overline{AF}$ to the extension of base $\overline{CB}$.

21. Measure the lengths of altitude $\overline{AF}$ and base $\overline{BC}$.
    Use these lengths to calculate the area of the triangle.

22. What do you notice about the area calculations? Explain what this shows.

23. Using a pencil, draw any bases and heights in the triangle below.

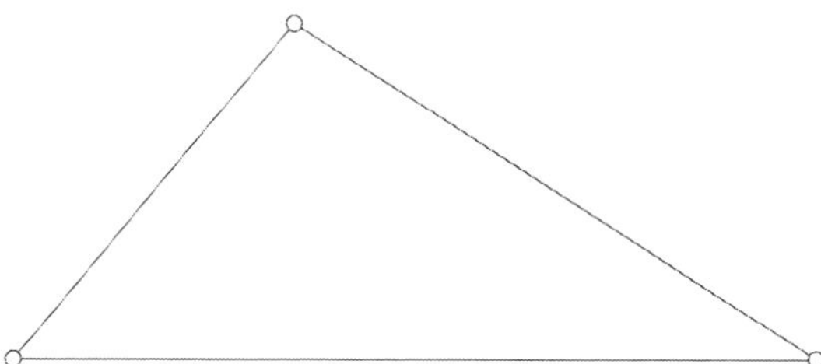

24. Compare your drawing in step 23 to your drawing in step 1. What have you learned?

EXPLORE MORE

25. As you make the Sketchpad triangle acute, obtuse, and right, watch the three altitudes. Write about your findings. Can you make a triangle in which exactly two of the altitudes fall inside the triangle? Explain.

26. Make your triangle obtuse and extend the three altitudes by constructing lines. What do you notice? The point where the altitudes intersect is called the *orthocenter* of the triangle. Write some observations about where the orthocenter is located. Consider obtuse, acute, and right triangles.

Sliding Bases: Parallelogram Area

ACTIVITY NOTES

INTRODUCE

Project the sketch for viewing by the class. Expect to spend about 5 minutes.

1. Explain that students will use what they know about rectangles to explore parallelograms and make some interesting discoveries.
2. Open **Sliding Bases.gsp.** Introduce any Sketchpad skills new to your students. Show that the parallelogram passes the drag test, remaining a parallelogram as sides and vertices are dragged. To demonstrate how Sketchpad can measure the area, select the interior of the parallelogram and choose **Measure | Area.**
3. If you want students to save their Sketchpad work, demonstrate choosing **File | Save As,** and let them know where to save.

DEVELOP

Expect students at computers to spend 30 minutes.

4. Assign students to computers and tell them where to locate **Sliding Bases.gsp.**
5. Distribute the worksheet. Tell students to work through the steps of the worksheet and do the Explore More if they have time. Encourage them to ask a neighbor for help if they have questions about using Sketchpad.
6. Let students work at their own pace. As you circulate, observe students at work, listen to their conversations, and ask guiding questions.
   - Students start by manipulating a parallelogram to get a feel for the properties of the figure. Ask, ***What stays the same as you drag vertices and edges?*** [Opposites sides remain parallel and equal in length.] ***What changes?*** [Shape (long and skinny versus short and fat), size, side length, perimeter, orientation, area]
   - Working on page "Grid" in worksheet step 3, students move their parallelogram until it forms a square with area 1 square unit. This reinforces the idea that square units are the unit of measure for area.
   - The area dynamically updates as students drag sides of the rectangle so that it has a base of 3 cm and area covering 6 squares, or 6 cm^2 (worksheet step 5). Students should be able to make the rectangle whether or not they figure out in advance that the rectangle must have a height of 2 centimeters.
   - The question in worksheet step 6 gives students a chance to express the connection between base, height, and area.

- Depending on the level of abstraction you feel your students are ready for, in worksheet step 8 you might encourage them to write the formula using letters for variables ($A = bh$).

7. As students work on worksheet step 10, they may focus on parallelograms that will fit on the screen. If students focus on only those parallelograms, ask, ***What if the parallelogram could be bigger than the screen? These parallelograms' vertices are all on grid lines, but what if the vertices didn't need to be on the grid points?***

8. Worksheet step 11 could be described as *shearing*. In the case of a parallelogram, shearing changes the shape of the parallelogram by moving one base while keeping it the same distance from the other base. You might introduce the term. ***In step 11 you are shearing the parallelogram.***

   What changes when a parallelogram is sheared? [The parallelogram's shape, side length, and perimeter] ***What stays the same?*** [The parallelogram's base, height, and area]

   You say the height stays the same. What does the term* height *mean? [Height is the distance between the two bases. Height is the length of the segment between and perpendicular to the two bases.] This segment is called the *altitude*.

   You might ask a student who gives a good definition of *height* to share it during the summarizing discussion.

9. As students will see in worksheet step 14, they need to distinguish between height and length of a side to avoid the common misperception that parallelogram area can be found by multiplying side lengths.

Explore More

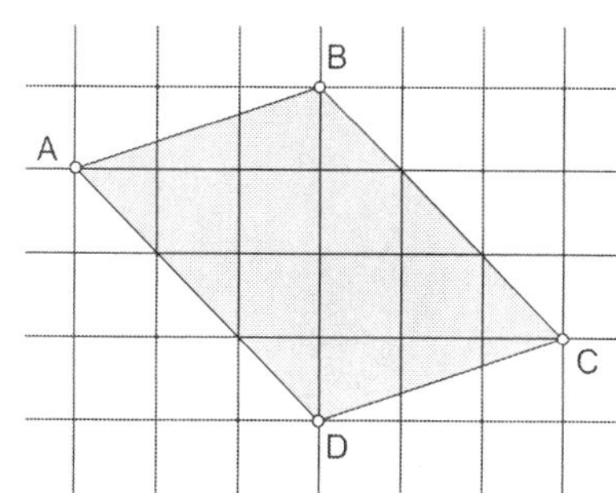

10. Students who complete the investigation quickly should be directed to the Explore More problem. If students don't discover that "tilted" parallelograms like the one shown here are possible, provide a hint. ***Does the base of the parallelogram have to be along a grid line? Do you think it's possible to make a tilted parallelogram that has an area of 12 square units?*** [Yes, many examples can be shown.]

SUMMARIZE

Project the sketch for viewing by the class. Expect to spend about 20 minutes.

11. Open **Sliding Bases.gsp** so it is available for illustrating students' ideas.
12. As students share answers to worksheet step 6, some may talk about dividing and others may state *Look for a number to multiply by 3 to get 6.* It is valuable for students to see different, but equally correct answers.
13. Begin a table on the board and ask students to share numbers from the tables they built in worksheet step 12.

    What relationships do you see between the numbers in the table?

    Does that relationship always hold? Explain. Encourage students to express their own impressions.

    What does* height *mean?

    Did anyone find a base and a height whose product was not the area?
14. Ask for answers to worksheet step 14. Discuss the explanations. Students should agree that you need the length of one side, but you also need the distance between the two sides of that length.

    How did you complete the formula in step 15, Area = ? Take responses and then write the formula on the board: *Area = base · height.*
15. Follow up with more open-ended questions that clarify the meaning of this formula and that check students' understanding.
    - ***How is finding the area of a parallelogram like finding the area of a rectangle?***
    - ***How is it different?*** [For a rectangle, base and height are side lengths; for a parallelogram, height is not a side length.]
    - ***Can we find the area of a parallelogram by multiplying side lengths? Why or why not?***
16. Provide practice in finding areas of a few parallelograms from a worksheet, from students' textbooks, or as drawn on the board. Include an example where only edge lengths are given.

Explore More

17. Ask students to share their answers to the Explore More challenge. Show some parallelograms whose sides are not along grid lines. One example is on page "Explore More". Ask them how they can tell the area is 12 cm^2. Look for explanations that include dissecting the parallelogram by moving triangles to make it a parallelogram with a horizontal base or a rectangle with the same area.

ANSWERS

2. It's always a parallelogram. Side $\overline{AB}$ is always parallel to $\overline{CD}$, and $\overline{AC}$ is parallel to $\overline{BD}$. The opposite sides are always equal in length.
4. Yes. (The measurement displayed will be 1 cm^2.)
5. The rectangle is 3 cm by 2 cm. The height is 2 cm.
6. Sample answer: The base tells you there are three squares in each row. The area will cover 6 squares, or 6 cm^2. You need 2 rows of 3; the height is 2 cm.
7. 8 cm^2
8. $Area = base \cdot height$
9. Yes. Sample answer: Dragging point C to the right or left does not change the base or the height, so it doesn't change the area.
10. Sample answer: Infinitely many (or an unlimited number of) parallelograms. If you drag point C so that that base and height stay unchanged, you can make infinitely many parallelograms. All the parallelograms with base 4 cm and area 8 cm^2 have a height of 2 cm.
11. The height is 3 cm.
12. Answers will vary. In students' tables, each area should be the product of the base and height.
13. The length of the base and the height do not change.
14. Sample answer: One side length is needed as the length of the base. The height is needed rather than the measure of the other side.
15. $Area = base \cdot height$
16. Answers will vary depending on computer platform.

Sliding Bases

Name:

Use what you know about rectangles to make discoveries about parallelograms.

EXPLORE

1. Open **Sliding Bases.gsp** and go to page "Quadrilateral."

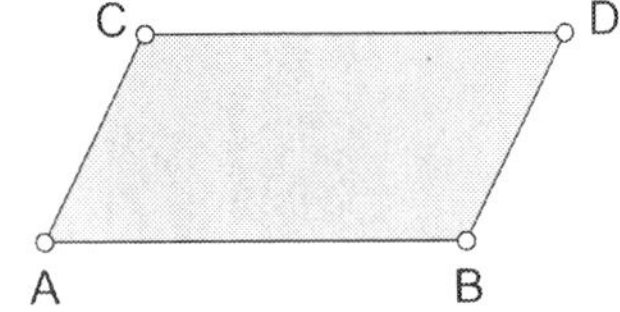

2. Drag edges and points *A*, *B*, *C*, and *D*. What do you notice about quadrilateral *ABCD*? What is another name for this shape?

3. Go to page "Grid."

   Drag points *A*, *B*, and *C* until your parallelogram is a square with area one square unit.

4. Now measure the parallelogram's area. Select the interior of the square. Choose **Measure | Area.**

   Did you confirm that the area is 1 cm^2? ______________________

5. Drag points *B* and *C* to make a rectangle with a base $\overline{AB}$ of 3 cm and area of 6 cm^2.

   Describe this rectangle. What is its height?

6. How does knowing the base and area of a rectangle let you find its height?

7. Make a rectangle with base 4 cm and height 2 cm. What is its area?

8. Use the words *base* and *height* to write the formula for the area of a rectangle.

   Area = ____

9. Can you drag point *C* so that the base $\overline{AB}$ stays 4 cm, and the area of the parallelogram stays 8 cm^2, even when *ABCD* is not a rectangle? Explain.

Sliding Bases

continued

10. How many parallelograms are there with base 4 cm and area 8 cm²? Explain.

    What is the height of these parallelograms? ____________________

11. Make the base of the parallelogram 5 cm and the area 15 cm². Drag point *C* so that you change the shape of the parallelogram, but not the base or area.

    What is the height? ______________________________

12. Now experiment with parallelograms of different areas. Record these areas, bases, and heights in the table.

Area	Base	Height

Area	Base	Height

13. For each area measure, drag point *C* so that the area doesn't change. What else doesn't change as you drag point *C*?

14. Is it useful to know the lengths of a parallelogram's sides when you are finding the area of a parallelogram? Explain.

15. Use the words *base* and *height* to write the formula for the area of a parallelogram.

    Area = ______________________________

EXPLORE MORE

16. Go to page "Explore More." Look for differently shaped parallelograms (including those shaped like rectangles) with area 12 cm² and with vertices on the grid intersections. How many parallelograms can fit within the square?

Shape Shearing: Parallelogram Area Formula

ACTIVITY NOTES

INTRODUCE

Project the sketch for viewing by the class. Expect to spend about 10 minutes.

1. Review students' understanding of parallelograms.

   Ask, ***What is a parallelogram?*** [Four sides, opposite sides parallel, opposite sides of equal length]

   How can you find the area of a parallelogram? [*Area* = *base* · *height*]. Write the formula on the board. ***The area of a parallelogram is the length of the base times the height.***

2. Open **Shape Shearing Present.gsp** and go to page "Parallelogram." Drag vertices or sides to show the quadrilateral remains a parallelogram. Ask, ***How could you find the height of this parallelogram?*** Bring out the idea that height is measured along a perpendicular between two parallel bases.

3. ***In this activity you'll start with this parallelogram and add a segment that represents its height. You'll then apply the formula Area = base · height to calculate its area. Finally, you'll find a way to change the shape of your parallelogram without changing its area.***

4. Model parts of the construction students will do, but leave some things for them to discover.

   - As you demonstrate constructing an altitude, worksheet steps 2–8, introduce or review other Sketchpad skills, as appropriate for your students.
   - Show students how to measure and how to enter a measurement into Sketchpad's Calculator, worksheet steps 10 and 11.

5. If you want students to save their Sketchpad work, demonstrate choosing **File | Save As,** and let them know where to save.

DEVELOP

Expect students at computers to spend 25 minutes.

6. Assign students to computers and tell them where to locate **Shape Shearing.gsp**.

7. Distribute the worksheet. Tell students to work through the worksheet and to do the Explore More if they have time. Let students know that they can ask others for help when they have questions about following the Sketchpad steps.

8. Circulate among students, asking questions and listening to the conversations.

   - As students work on steps 3, 4, and 5, ask, ***What happens when you drag vertices or sides? What changes? What stays the same? Does the construction hold together?*** [Dragging will reveal a mistake in step 4 for students who draw a line that *appeared* to go through the two points but missed at least one of them.]
   - At worksheet step 14 ask, ***What have you tried?***
   - ***Which vertex can you drag to change the shape of the parallelogram without changing the length of the base?***
   - ***Can you move a vertex and keep the height about the same?***

9. Worksheet step 14 anticipates steps 15–18, in which students constrain the figure so that it can be dragged without changing the base or height. Check in with pairs; ask them to share their thinking.

10. Asks pairs about their explanations for worksheet step 21. Find students whose answers you will ask for during the summarizing discussion.

11. Look for a pair who has completed the Explore More to share their work with the class. If they are unable to save it on the classroom computer, they might add a page to your sketch with the construction as other students are finishing the activity.

SUMMARIZE

Expect to spend 10 minutes.

12. Go to page "Shearing." Bring the class together to discuss the questions on the worksheet and any insights or misconceptions you observed as students worked. Ask additional questions such as, these.

    Why is the base $\overline{AB}$? Could the base be side $\overline{CD}$?

    Why did height $\overline{DE}$ have to be perpendicular to base $\overline{AB}$? [The height is the distance between the bases; *distance* means the shortest distance, so it has to be along a perpendicular.]

13. ***Why does the area stay the same when a parallelogram is sheared?*** [The altitude was constrained to the parallel line; the parallelogram had a constant area as long as the base and height remained the same.] Be sure students observe that the base and height are kept constant in each case. You might ask a few students to share their sketches. ***What is the base in this example? The height?***

14. Discuss extreme cases some students are likely to have explored—very stretched-out parallelograms with large perimeters. Ask students to explain how a parallelogram can have a large perimeter without having a large area. Reinforce the idea that slant height does not affect area.

15. ***What other questions could be explored?***

    Are there other ways to measure the area of a parallelogram? Could you use the lengths of the diagonals? Could you use the measures of the angles?

    What is the parallelogram with the smallest perimeter for a given area?

    Is the area the same no matter which side you use as the base?

16. Discuss the Explore More activity. Demonstrate or have a student demonstrate that any side can be a base.

ANSWERS

9. Sample answer: This segment represents the height. [Students may also include the words *altitude, measurement,* or *distance.*]

12. The calculation is $m\overline{AB} \cdot m\overline{DE}$, the measure of this parallelogram's base times the measure of its altitude (the height).

14. If you can drag point D without changing the base or height, the area won't change.

19. Sample answer: The parallelogram can be slanted and made long and skinny, but base $\overline{AB}$ and height $\overline{DE}$ don't change in length. [Or, the measures of the base and the height don't change.]

21. Sample answer: Dragging point D changes the lengths of two sides of the parallelogram and thus the perimeter. But it doesn't change the length of the base or the height. Because the area depends only on these lengths, the area doesn't change.

22. Make the parallelogram long and skinny.

23. A square.

24. Finding the area using this new altitude and base $\overline{BC}$ resulted in the same area measurement.

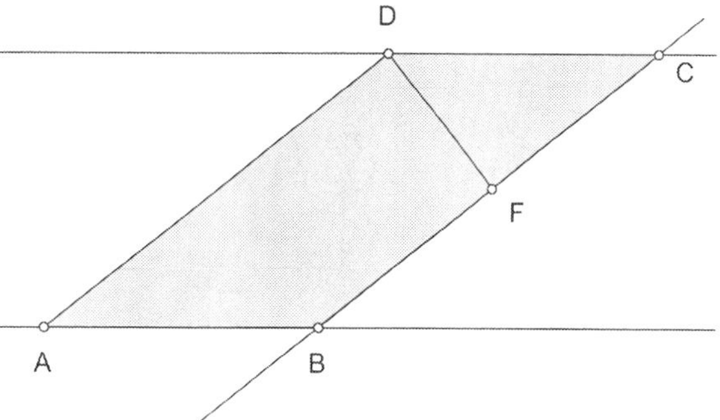

Shape Shearing

Name:

In this activity you will apply what you know about parallelogram area. Then you'll learn to change the shape of a parallelogram while keeping the area constant.

EXPLORE

1. Open **Shape Shearing.gsp** and go to page "Parallelogram."
2. Select the interior and choose **Measure | Area.**
3. Drag points *A*, *B*, and *C*, and experiment with how changing the shape changes the area.

CONSTRUCT

You can calculate the area of a parallelogram if you know the lengths of its base and its height. Now you'll construct an *altitude*, a segment whose length is the height of the parallelogram.

4. Construct a line through points *A* and *B*.

5. Drag point *D* so that it is not over the side $\overline{AB}$.
6. Select point *D* and $\overleftrightarrow{AB}$; then choose **Construct | Perpendicular Line.**

7. Construct a point at the intersection of $\overleftrightarrow{AB}$ and the perpendicular line.
   Label the new point *E*.

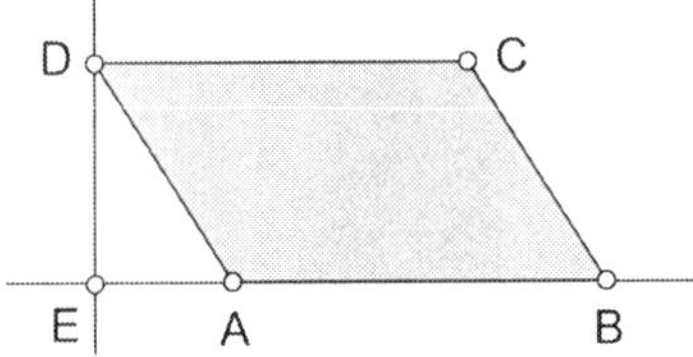

8. Select $\overleftrightarrow{DE}$; and choose **Display | Hide Perpendicular Line.**

   Construct $\overline{DE}$ (a segment) in its place.

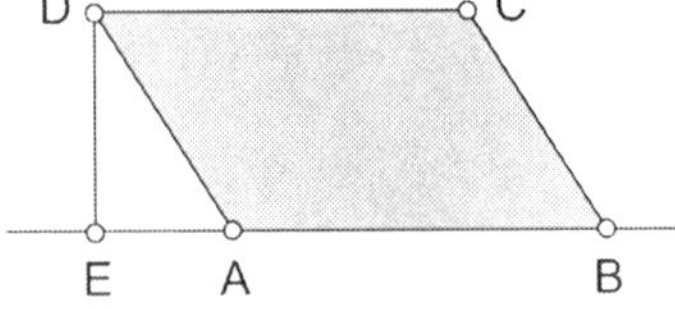

9. What does $\overline{DE}$ represent for the parallelogram?

10. Select two segments—the base $\overline{AB}$ and the height $\overline{DE}$—and choose **Measure | Length.**

EXPLORE

11. Choose **Number | Calculate.** Calculate the area of the parallelogram (click on the measurements in your sketch to include them in your calculation).

12. Write the calculation you used.

13. Drag points *A*, *B*, and *C* to confirm that your calculation always equals the measured area.

14. Describe how you can change the shape of the parallelogram without changing its area.

CONSTRUCT

In steps 15–19, you will explore a way to give your parallelogram constant area.

15. Construct a point *F* above the parallelogram.

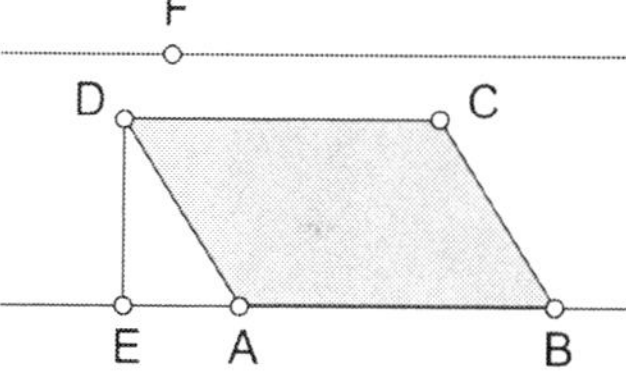

16. Select point *F* and $\overline{AB}$; then choose **Construct | Parallel Line.**

17. Select point *F* and choose **Display | Hide Point.**

18. Select point *D* and the parallel line and choose **Edit | Merge Point to Line.**

EXPLORE

19. Drag point *D*. How does the parallelogram change?

    What doesn't change in the parallelogram?

20. Select the parallelogram interior and choose **Measure | Perimeter.**

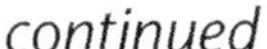

21. Explain why dragging point *D* changes the perimeter of the parallelogram but not the area.

22. Keeping the area constant, how can you make the perimeter very large?

23. What kind of parallelogram has the least perimeter for a given area?

EXPLORE MORE

24. Is $\overline{AB}$ the only side that can be a base? To explore this question, construct an altitude $\overline{DF}$ from point *D* to a line through side $\overline{BC}$. Use this new altitude and new base $\overline{BC}$ to calculate the area of the parallelogram.
    What did you find? Write and draw to explain your conclusions.

INTRODUCE

Project the sketch for viewing by the class. Expect to spend about 10 minutes.

1. Open Sketchpad and enlarge the document window so it fills most of the screen.
2. Explain, ***Today you're going to use Sketchpad to discover the area of a trapezoid. First let's define* trapezoid. *What is a trapezoid?*** Listen for the following responses.

   A trapezoid is a plane figure; it's a polygon.

   A trapezoid has four sides; it's a quadrilateral.

   A trapezoid has two sides that are parallel.

   The parallel sides are called bases.

   The two nonparallel sides are called legs.

   Is a parallelogram a trapezoid? Explain. [Yes, if you use an inclusive definition of trapezoid (having at *least* one pair of parallel sides); no, if you use an exclusive definition of trapezoid (having *exactly* one pair of parallel sides).] Write a definition for *trapezoid* on chart paper. Here is a sample definition: A trapezoid is a quadrilateral that has one pair of parallel sides. The parallel sides are called *bases* and the nonparallel sides are called *legs.*
3. ***I'll demonstrate how to construct a trapezoid using the definition we wrote.*** As you demonstrate, make lines thick and labels large for visibility. Model the trapezoid construction in worksheet steps 1–8. Here are some tips.
   - Start by choosing **Edit | Preferences | Text** and choosing **Show labels automatically: For all new points.** Explain that for purposes of the demonstration, points will be automatically labeled as you construct the trapezoid. Tell students that they will label their points using the **Text** tool after they construct the trapezoid.
   - In worksheet steps 1 and 2, talk about the parts of the trapezoid as you construct them. ***Segment AB will be one of the bases of the trapezoid. Segment AC will be one of the legs. What will I need to do next to construct the other base?*** [Make a line parallel to $\overline{AB}$ through point *C*.] Construct the parallel line.
   - In worksheet step 4, ask students whether it matters where point *D* is placed. ***According to our definition, does it matter where I construct***

point D on the line? [Yes, point *D* needs to be to the right of point *C* for the shape to have four sides.]

- Follow worksheet steps 6–8 to finish constructing the trapezoid. ***How does this shape meet our definition of a trapezoid?*** [The shape has four sides and one pair of parallel sides, $\overline{AB}$ and $\overline{CD}$.] ***What is the name of this shape?*** [Trapezoid *ABDC*] Remind students to name a trapezoid by listing the vertices in consecutive order.
- Model how to choose **Edit | Undo.** Explain that students can use this command to "undo" a previous action if they make a mistake in their construction.

4. If you want students to save their work, demonstrate choosing **File | Save As,** and let them know how to name and where to save their files.

DEVELOP

Expect students at computers to spend about 20 minutes.

5. Assign students to computers. Distribute the worksheet. Tell students to work through step 20 and do the Explore More if they have time. Encourage students to ask their neighbors for help if they are having difficulty with the construction.
6. Let pairs work at their own pace. As you circulate, here are some things to notice.
   - In worksheet step 11, check students' understanding of the term *height.* ***Can you explain what the height of the trapezoid is?*** [It is the perpendicular distance from one base to the other base.] Explain that when Sketchpad measures a distance between a point and a line, it measures the shortest distance, or the length of a perpendicular line segment from the point to the line. For your understanding, height is also referred to as *altitude.*
   - In worksheet step 12, listen to students as they drag different parts of the trapezoid. Ask questions to check their understanding. ***What happens to your measurements when you drag point D to the right?*** [The base $\overline{CD}$ gets longer, the base $\overline{AB}$ and the height stay the same, and the area gets larger.] ***What happens when you drag $\overline{CD}$ up?*** [The lengths of the bases stay the same, while the height and area increase.] At this point, students only observe that changing the length of the base changes the area and changing the height changes the area.

- In worksheet step 14, tell students that point E will flash briefly to indicate that it has been marked as a center. ***This point will be the center of the rotation.*** Have students try to guess what shape the trapezoid and its rotated image will form before they rotate the trapezoid. For your understanding, students will need to select points A and C as well as the interior of the trapezoid, so that points A' and C' appear in the rotated image. Remind students that in the rotated shape, the corresponding point to A is called "A prime" and is written A'. The corresponding point to C is C'.
- In worksheet step 16, students should recognize the new shape as a parallelogram. Have students identify the parts of the parallelogram. ***What is the base of the new shape?*** [The length of $\overline{AC'}$ or length of $\overline{CA'}$] ***Is*** $\overline{DB}$ ***the height of the new shape?*** [No] ***How do you know?*** [It is not perpendicular to the base.] Note that the height needs to be constructed.
- In worksheet step 17, help students as needed. ***What is the length of*** $\overline{BC'}$ ***equal to?*** [It is the same length as $\overline{CD}$.]
- In worksheet step 18, students will need to recall the formula for the area of a parallelogram. ***What is the area formula for the new shape?*** [$A = bh$]
- Check that students use parentheses around the sum of the bases. If not, try to get them to recognize their error. ***Is*** $4 + 6 \div 2$ ***the same as*** $(4 + 6) \div 2$***? Explain.*** [No; following the order of operations, the value of the first expression is 7 and the value of the second expression is 5.]
- If students have time for the Explore More, they will derive another formula for the area of a trapezoid using the median.

7. If students will save their work, remind them where to save it now.

SUMMARIZE

Project the sketch. Expect to spend about 15 minutes.

8. Gather the class. Students should have their worksheets with them. Open **One Parallel Pair Present.gsp** and go to page "Trapezoid." Begin the discussion by having students identify the new shape formed by the transformation. ***What new shape was formed by the trapezoid and its rotated image?*** [Parallelogram $AC'A'C$] ***How do you know?*** [The shape is a parallelogram because opposite sides are parallel.]

9. Have students explain how they arrived at their answers to worksheet step 17. ***What is the base of the new shape if b_1 represents the length of $\overline{AB}$ and b_2 represents the length of $\overline{CD}$? Explain your reasoning.*** Students should reason that $\overline{CD}$ and $\overline{BC'}$ are the same length; $\overline{BC'}$ is just the rotated image of $\overline{CD}$. The base of the parallelogram is $m\overline{AB} + m\overline{BC'}$ or $m\overline{AB} + m\overline{CD}$ or $b_1 + b_2$.

10. For worksheet step 18, ask volunteers to share their answers. ***What area formula did you use?*** [Area of a parallelogram, $A = (b_1 + b_2)h$] ***Why are the parentheses important in your formula?*** [You need to add the two base measurements before multiplying by the height.]

11. For worksheet step 19, ask students for the formula of a trapezoid. ***How did you arrive at your formula?*** Students' reasoning may vary. Here are some possible replies.

    It took two trapezoids to make the parallelogram, so the area of one trapezoid is one-half the area of the parallelogram. So it is $bh \div 2$, *where b is the sum of the two bases of the trapezoid, or* $(b_1 + b_2)h \div 2$.

    The heights of the trapezoid and the parallelogram are the same. The base of the parallelogram is equal to both bases of the trapezoid added together. So we substituted $b_1 + b_2$ *for b in the area formula for a parallelogram and got* $h(b_1 + b_2)$. *That's the area of the parallelogram. But two trapezoids formed the parallelogram, so to find the area of one trapezoid, we divided by 2 and got* $h(b_1 + b_2) \div 2$.

12. Review worksheet step 20 by pressing *Show Calculation* on the sketch. Have students verify that their expressions match. Point out that the value of the expression and the area of trapezoid *ABDC* are equal.

13. Go to page "Double Trapezoid" and press *Double Trapezoid* to show another example of the relationship between the area of a trapezoid and the area of a parallelogram formed by the trapezoid and its rotated image. It is clear in this sketch that the new shape has a base measurement of $b_1 + b_2$, which are the two bases of the trapezoid.

14. If time permits, discuss the Explore More. ***How is the midsegment related to the two bases?*** Students should understand that the midsegment (median) of a trapezoid is also the average length of the two bases, so the height times the midsegment (median) length is equal to the area.

ANSWERS

16. The two combined trapezoids form a parallelogram.
17. The parallelogram formed by the combined trapezoids has a base of length $b_1 + b_2$.
18. $A = (b_1 + b_2)h$ is the area of the parallelogram.
19. $A = \frac{1}{2}(b_1 + b_2)h$ is the area of the trapezoid.
20. $\dfrac{(m\overline{AB} + m\overline{CD})(Distance\ C\ to\ \overline{AB})}{2}$

    Students must use parentheses when they sum the two bases.
21. $A = mh$, where m is the length of the midsegment, gives the area of the trapezoid. The length of the midsegment is the average of the lengths of the bases.

One Parallel Pair

Name:

In this activity you'll construct a trapezoid and then transform it into a familiar shape. You'll use the formula for the area of this shape to write the formula for the area of a trapezoid.

CONSTRUCT

1. Construct two segments that share an endpoint.

2. Label the shared endpoint *A* and the other two endpoints *B* and *C*.

3. Now you'll construct a line through point *C* parallel to $\overline{AB}$.
   Select point *C* and $\overline{AB}$ and choose **Construct | Parallel Line.**

4. Construct point *D* on the line and label it.

5. Construct $\overline{BD}$.

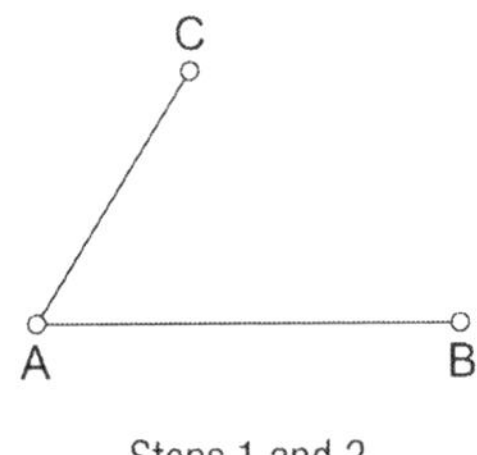

Steps 1 and 2

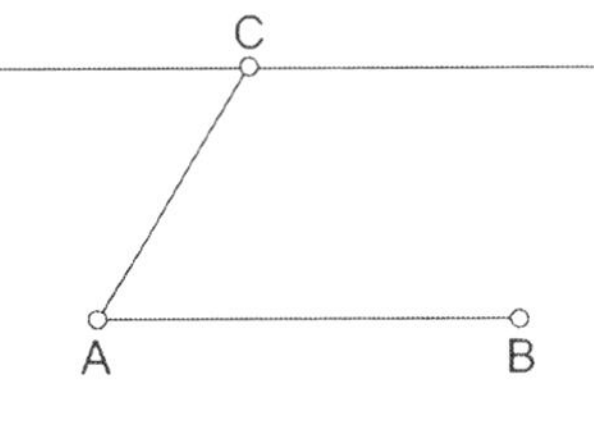

Step 3

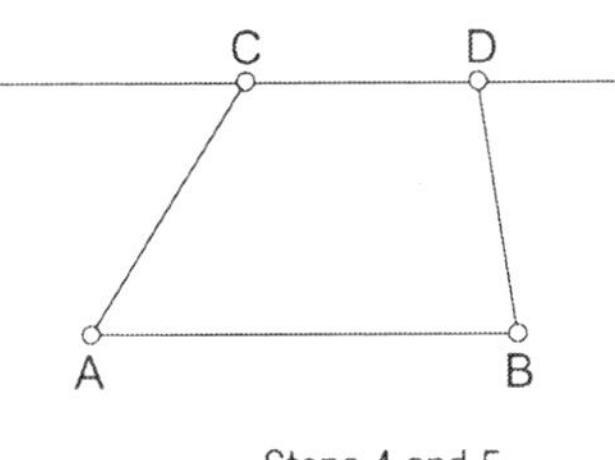

Steps 4 and 5

6. Hide the line by selecting the line and choosing **Display | Hide Parallel Line.**

7. Construct $\overline{CD}$.

8. Next, you'll construct the interior of trapezoid *ABDC*.
   Select the vertices in consecutive order and then choose **Construct | Quadrilateral Interior.**

EXPLORE

Area ABDC = 4.20 cm^2
m $\overline{AB}$ = 3.78 cm
m $\overline{CD}$ = 2.15 cm
Distance C to $\overline{AB}$ = 1.42 cm
C D
A B

9. Measure the area of *ABDC* by selecting the interior and choosing **Measure | Area.**

10. You will measure the lengths of the bases of the trapezoid, $\overline{AB}$ and $\overline{CD}$.
    Select each base and choose **Measure | Length.**

11. Now you'll measure the height of the trapezoid.
    Select point *C* and $\overline{AB}$. Then choose **Measure | Distance.**

12. Drag different parts of the trapezoid and observe the measures.

At this point, it's probably hard to see any relationships between the area measure and the base and height measurements.

CONSTRUCT

13. Select $\overline{DB}$ and choose **Construct | Midpoint.**
    Label the midpoint *E*.

14. Now you'll mark point *E* as a center of rotation and rotate the entire trapezoid by 180°.
    Double-click point *E* to mark it as a center.
    Select points *A* and *C* and then the interior of the trapezoid.
    Choose **Transform | Rotate.**
    In the dialog box, enter 180 for Angle.
    Click **Rotate.**

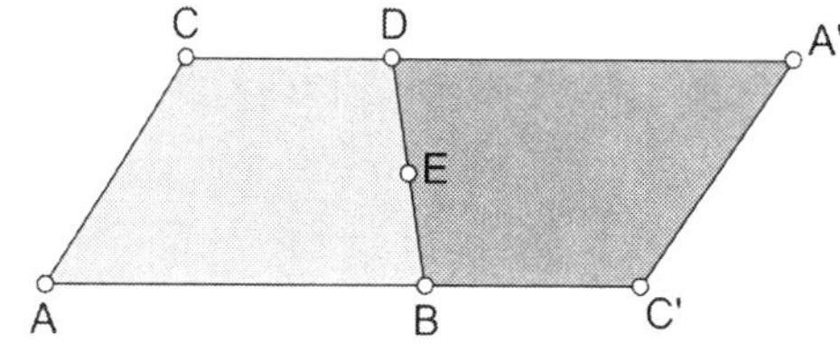

EXPLORE

15. Drag parts of the figure and observe the shape formed by the trapezoid and its rotated image.

16. What shape do the two combined trapezoids form?

17. Let b_1 represent the length of base $\overline{AB}$ and let b_2 represent the length of base $\overline{CD}$. What is the length of the base of the shape formed by the combined trapezoids?

18. Write a formula for the area of the combined shape in terms of b_1, b_2, and h (for height).

One Parallel Pair

continued

19. Write a formula for the area of a single trapezoid in terms of b_1, b_2, and h.

20. In your sketch, check that you've derived the correct formula by calculating an expression equal to the area of the trapezoid. Use $m\overline{AB}$, $m\overline{CD}$, and the distance from point C to $\overline{AB}$ in your expression.

    Choose **Number | Calculate** to open the Sketchpad Calculator.

    Click once on a measurement to enter it into a calculation. Use parentheses where necessary. Record your expression.

EXPLORE MORE

21. Construct the midpoints of the nonparallel sides of the trapezoid. Connect these midpoints with a segment. Use the length of this midsegment to invent a new area formula.

If you are only doing a short demonstration after students have already derived the formula for circle area, show the model, and then begin your discussion with step 6.

INTRODUCE

Project the sketch for viewing by the class. Expect to spend about 20 minutes.

1. Open **Wavy Parallelogram.gsp** and go to page "Circle Sectors." Ask students to study and then describe the two representations. ***How are the circle and this "wavy parallelogram" shape related?*** [Each is made up of the same number of circle sectors of the same size.]
2. Give students plenty of time to study the representations, and then begin discussion. Students are likely to recognize that the sectors of the circle have been rearranged in a way that approximates a parallelogram. (Inside the wavy parallelogram is the outline of an actual parallelogram.) ***What can you say about the area of these figures?*** Elicit the idea that the area of the two figures must be the same.
3. Ask the following two questions, and then use a think-pair-share approach to facilitate students' making sense themselves of the relationships between the parts of the wavy parallelogram and the parts of the circle.

   As you trace along the wavy base, say, ***Look at the wavy base of the wavy parallelogram. How is it related to the circle?*** [It's half the circumference.]

   How is the height of the wavy parallelogram related to the circle? [It's the radius.]

   The discussion should bring out the ideas that the base of the wavy parallelogram is half the circle's circumference and that the height is the circle's radius.

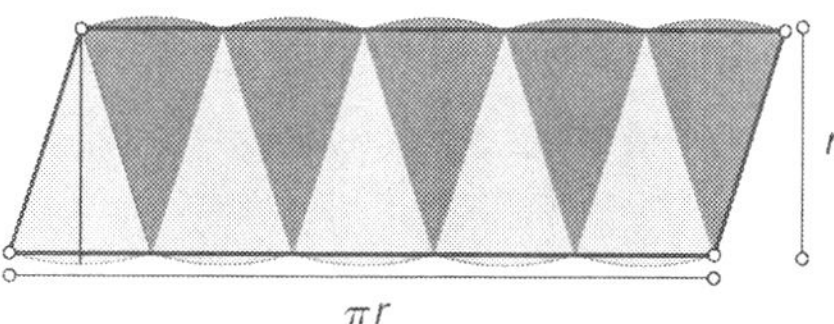

4. Drag the slider slowly, pausing at 6, 8, and 10 and stopping at 12. Students will observe that as the number of sectors increases, the wavy parallelogram more closely approximates a parallelogram. ***You observed that the area of the circle and the area of the wavy parallelogram are the same. How can you use what you know about parallelogram area***

to find the area of the circle? Work together to figure this out. Allow plenty of time for students to work in pairs or small groups and struggle with this question. If you need to prompt some students, ask questions such as these.

How can knowing the formula for the area of a parallelogram help you? Make sure students recall the formula.

The base of the wavy parallelogram is half the circumference of the circle. How can you use what you know about the formula for the circumference of a circle, $C = \pi d$, or $C = 2\pi r$?

5. When students are ready, bring the class together again and invite students to share their thinking. ***Let's hear your ideas about finding the area of the circle.*** The discussion should bring out the following ideas.
   - Using r for radius of the circle, the circumference of a circle is equal to $2\pi r$. Because the base of the parallelogram is half of the circumference, it is equal to πr.
   - The height of the parallelogram is equal to the radius of the circle, r.
   - The formula for area of a parallelogram is $A = b \cdot h$.
   - We can substitute πr for the base, and r for the height.
   - Multiplying base times height in the wavy parallelogram gives $\pi r \cdot r$, or πr^2.
   - Let students know that this new formula is a formula for circle area.

DEVELOP

Expect to spend about 15 minutes. Continue to project the sketch.

6. ***Are you convinced that using the formula for the area of a regular parallelogram in the way we did works for a wavy parallelogram? Were any of you concerned that a wavy parallelogram isn't a "real" parallelogram?*** Give students a moment to express any uncertainty. Acknowledge their responses and express respect for their thoughtfulness.

   Sketchpad gives us a way to look a little deeper into why your formula works. We can drag the slider more and that will increase the number of sectors. How will increasing the number of sectors affect how wavy the parallelogram is? Students are likely to intuit correctly that fewer sectors make wavier parallelograms and that more sectors result in a straighter base. Drag the slider all the way to 18 to illustrate this.

7. Drag the slider back to 6 and direct students' attention to the actual parallelogram constructed within the wavy parallelogram. ***Why is the area of the actual parallelogram less than the area of the circle?*** Students should observe that wavy parts of the circle extend outside the parallelogram because the radius of the circle is greater than the height of the parallelogram.

8. ***What do you think will happen to the areas of the parallelogram and circle as I increase the number of sectors again? Why?*** Students predict that the areas will get closer to one another—specifically, the area of the parallelogram will increase while the area of the circle stays constant. The parallelogram area increases because the height gets closer to the radius of the circle. Drag the slider again, slowly, and discuss students' observations and thinking. Follow up with three more questions.

   What do you notice about the base of the actual parallelogram compared to the base of the wavy one as the number of sectors increases? [The base of the wavy parallelogram more closely approximates the base of the actual parallelogram.] ***What if we had 100 sectors? A 1000?***

   What do you notice about the sides of the parallelogram as the number of sectors increases? [The sides make a larger angle with the base as the number of sectors increases.] ***What would the sides be like with a very large number of sectors?***

   Compare the height of the actual parallelogram and the height of the wavy one. [The height of the wavy parallelogram more closely approximates the height of the actual parallelogram as the number of sectors increases.] ***How would those heights compare if there were a very large number of sectors?***

SUMMARIZE

Expect to spend about 10 minutes. Continue to project the sketch.

9. ***How does the Sketchpad demonstration help confirm that we really can use parallelogram area to find the area of a circle?***

   The more sectors the circle is cut into, the more closely the circle approximates a parallelogram.

   If we could cut a circle into a very large number of sectors, we would get a parallelogram.

If we could cut a circle into infinitely many sectors, we would get a rectangle.

Let students use their own form of inductive reasoning to extend their thinking to a very large number of sectors.

There is no limit to how close we can get to a parallelogram.

We get so close that the parallelogram becomes a rectangle.

We can derive the formula for the area of a circle from the formula for the area of a rectangle.

10. Suggest that the class write an explanation for deriving the formula for the area of a circle. You might act as scribe, recording the explanation and asking for clarification as needed in order to encourage clear communication. Alternatively, have students write individually in response to the following prompt.

    How were you able to use what you know about parallelogram area to come up with a formula for circle area?

EXTEND

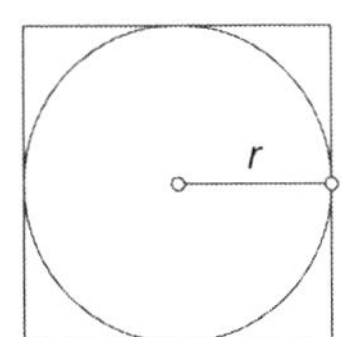

1. Show page "Circle in Square," a circle and a square around it whose side midpoints just touch the circle, as shown. ***What is the area of this square?*** [$A = 4r^2$] You might challenge students to construct a Sketchpad circle and square that will always have this relationship.

Smoothing the Sides: Regular Polygon and Circle Area

INTRODUCE

Project the sketch for viewing by the class. Expect to spend about 10 minutes.

1. Open **Smoothing the Sides.gsp** and go to page "Polygon." Enlarge the document window so it fills most of the screen.

2. Explain, ***Today you're going to use Sketchpad to explore the areas of regular polygons. Based on your investigations, you'll write area formulas for any regular polygon, and then for a circle. What is a regular polygon?*** Work with students to come up with a common definition. Here is a sample definition: A regular polygon is a polygon that has all congruent sides and all congruent angles.

3. ***What is the name of this regular polygon?*** [Regular hexagon] Remind students, if needed, that polygons are named by the number of sides. Review the names of common regular polygons.

Number of Sides	Regular Polygon
3	equilateral triangle
4	square
5	regular pentagon
6	regular hexagon
7	regular heptagon
8	regular octagon
9	regular nonagon
10	regular decagon
12	regular dodecagon

4. ***Suppose I draw a segment from the center of a regular polygon to each vertex. What shapes would I form?*** [Triangles] Use the **Arrow** tool to select the segments with a thick outline on the sketch. ***Into how many triangles is this regular hexagon divided?*** [Six] ***Are these congruent triangles? How do you know?*** [Yes, each has sides of the same length.] ***Into how many congruent triangles do you think a square can be divided?*** [Four] Let students guess, and then drag the marker to 4 to confirm. Explain that by dragging the marker you can change the number of sides of the regular polygon. Let students explore this relationship further on their own.

5. Review what *s* and *a* represent in the sketch. ***What does s represent?*** [Side length] ***What does a represent?*** [Apothem] Students may say that

this is the height of the triangle, which is true. Explain that it is also an *apothem* of the regular polygon, and provide a definition. Here is a sample definition: An apothem of a regular polygon is the line segment from the center to the midpoint of a side.

6. Tell students that they will now explore the sketch on their own to write area formulas for any regular polygon and for a circle.

DEVELOP

Expect students at computers to spend about 25 minutes.

7. Assign students to computers and tell them where to locate **Smoothing the Sides.gsp.** Distribute the worksheet. Tell students to work through step 14 and do the Explore More if they have time. Encourage students to ask their neighbors for help if they are having difficulty with the exploration.

8. Let pairs work at their own pace. As you circulate, here are some things to notice.

   - In worksheet step 4, students should already know that the formula for area of a triangle is $\frac{1}{2}bh$. Be alert to students who don't; they will need a quick review.
   - In worksheet steps 4 through 8, have students use the Sketchpad Calculator to verify their formulas. ***How can you test that your formula is correct?***
   - For worksheet steps 10 and 11, ask students who need help to observe the relationships among the measurements as the number of sides increases. ***Are any measurements almost the same?***
   - In worksheet step 13, have students who are having trouble review their answer for worksheet step 8. ***What formula did you write for the area of the regular polygon, using its apothem and perimeter? How do these measurements compare to the radius and the circumference of the circle?***
   - If students have time for the Explore More, they will investigate whether the ratio of the regular polygon's perimeter to twice its apothem approaches a familiar number, *pi*. Then students will write this ratio using the circumference and the diameter of the circle.

SUMMARIZE

Project the sketch. Expect to spend about 10 minutes.

9. Gather the class. Students should have their worksheets with them. Begin the discussion by reviewing worksheet step 4. ***What does "a" represent in the triangle?*** [Height] ***What does "s" represent in the triangle?*** [Base] ***What is the formula for the area of the triangle?*** $\left[\frac{1}{2}bh\right]$ ***What is the formula using "a" and "s"?*** $\left[\frac{1}{2}sa\right]$

10. Review worksheet steps 5 and 6. ***How many triangles are formed?*** [4 in the square; 5 in the pentagon] ***Are they all congruent?*** [Yes] ***How can you use this to find the area of each regular polygon?*** [Multiply the area of one triangle by the number of sides] Make sure students see that joining every vertex of any regular polygon to the center point forms as many isosceles triangles as there are sides of the polygon.

11. Now review worksheet step 7. ***What is the formula for the area of a regular polygon with n sides? Explain your reasoning.*** $[A = \frac{1}{2}asn]$ Students may come up with the following responses.

    The number of sides of a regular polygon is the same as the number of congruent triangles that make up the regular polygon. So you multiply the area of one triangle by the number of sides to find the total area.

    We found the area by first finding the total length of all the bases of the triangles. We did this by multiplying the base (or side length) by the number of sides, n. Then we multiplied by the height, or apothem, and divided by two.

    We discovered that you can construct n congruent triangles in a regular polygon with n sides. So, the formula for a regular polygon is the area of one triangle multiplied by n, the number of sides.

12. Review worksheet step 8, asking students to explain their formulas. ***How is the perimeter related to the variable s?*** $[P = sn]$ Make sure students understand that the variable P is being substituted into the area formula. This is necessary to derive the circle formula.

13. For worksheet steps 9–12, drag the marker to increase the number of sides of the regular polygon. ***What does the regular polygon look like now?*** [When the number of sides increases to 18 or greater, the regular polygon resembles a circle.] ***What does the apothem approach?*** [Radius] ***How about the perimeter?*** [Circumference] As the measurements get closer to each other, have students read them aloud.

14. For worksheet step 13, ask students to explain their answers. Students may reply with the following explanations.

    In step 8, we wrote the area formula for the regular polygon as one-half the apothem multiplied by the perimeter. As the regular polygon approaches a circle, the apothem approaches the radius and the perimeter approaches the circumference. We substituted the radius for the apothem and the circumference for the perimeter and came up with one-half the radius multiplied by the circumference.

    As the number of sides increases, the area of the regular polygon gets closer and closer to the area of the circle, the perimeter of the regular polygon gets closer and closer to the circumference of the circle, and the apothem gets closer and closer to the radius. We figured out that the area of the regular polygon is one-half its apothem multiplied by its perimeter, so the area of the circle is one-half the radius multiplied by its circumference.

15. In worksheet step 14, have students give the area for a circle after substituting and simplifying.

16. If time permits, discuss the Explore More. ***If the apothem approaches the radius as the number of sides increases, what does twice the apothem approach?*** [Diameter] ***What does the ratio of the perimeter to twice the apothem approach?*** [π] ***How can you write this ratio using the circumference and the diameter of the circle?*** [C/d]

ANSWERS

4. $A = \frac{as}{2}$ or $A = \frac{1}{2}as$
5. $A = 4\left(\frac{as}{2}\right)$ or $A = 4\left(\frac{1}{2}as\right)$
6. $A = 5\left(\frac{as}{2}\right)$ or $A = 5\left(\frac{1}{2}as\right)$
7. $A = n\left(\frac{as}{2}\right)$ or $A = \frac{1}{2}asn$
8. The perimeter is equal to s times the number of sides: $P = sn$

    $A = \frac{1}{2}aP$
9. As the number of sides increases, the polygon becomes more and more like a circle.
10. The apothem approaches the radius of the circle.
11. The perimeter approaches the circumference of the circle.

12. The area of the polygon approaches the area of the circle.

13. $A = \frac{1}{2}rC$

14. $A = \frac{1}{2}r(2\pi r)$

    $A = \pi r^2$

15. The ratio approaches the measurement for *pi,* 3.14 . . . .

16. $\frac{C}{d}$

Smoothing the Sides

Name:

In this activity you'll figure out how to calculate the area of any regular polygon by dividing it into congruent isosceles triangles. Then you'll explore how you can use this approach to calculate the area of a circle.

POLYGON AREA

1. Open **Smoothing the Sides.gsp** and go to page "Polygon."

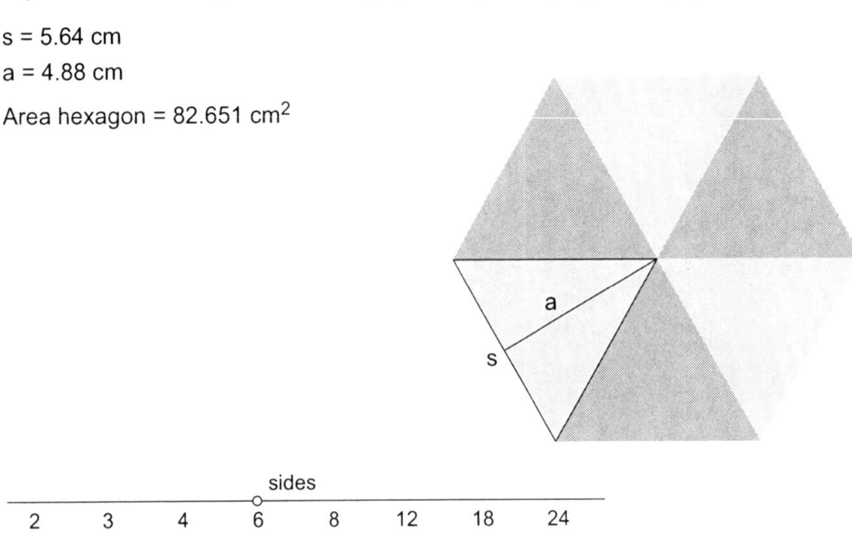

2. Drag the *sides* marker along its segment. Observe how the polygon and the measurements change.

3. Each regular polygon with four or more sides can be divided into triangles. Look at the triangle with a thick outline. It has segments labeled *a* for *apothem* and *s* for *side*. The side is the base of this triangle, and the apothem is the height of the triangle.

4. Go to page "Triangle," where you'll find an isosceles triangle. Choose **Number | Calculate** and write an expression to find the area of the triangle using *s* and *a*. Click on the values in the sketch to enter them into the Calculator. Check your expression against the measured area. Drag the vertices of the triangle to make sure your expression remains equal to the measured area for any isosceles triangle. What is your expression?

5. Go to page "Square." Press *Show Triangles* and then repeat step 4 to find the area of any square. What is your expression?

6. Go to page "Pentagon" and repeat step 5 to find the area of any regular pentagon. What is your expression?

7. Write an expression for the area of any regular polygon. Use n to represent the number of sides, as well as the variables s for the side length and a for the apothem length.

8. How is the perimeter of the polygon related to the variable s? Write another expression for the area of any regular polygon with perimeter P and apothem length a.

CIRCLE AREA

s = 5.64 cm
a = 4.88 cm
Area hexagon = 82.651 cm^2
Area circle = 99.94 cm^2
r = 5.64 cm
Circumference = 35.44 cm

a
s
r

sides
2 3 4 6 8 12 18 24

9. Go to page "Circle." Drag the *sides* marker. What happens to the polygon as the number of sides increases?

Smoothing the Sides

continued

10. What measure of the circle does the apothem approach as the number of sides increases?

11. What measure of the circle does the perimeter approach as the number of sides increases?

12. How does the measured area of the polygon compare to the measured area of the circle as the number of sides increases?

13. Write a formula for the area of a circle, similar to the formula you wrote in step 8, using C for circumference and r for radius.

14. The formula for circumference is $C = 2\pi r$. Substitute $2\pi r$ for C in your formula in step 13 and simplify.

EXPLORE MORE

15. Use the sketch to explore the ratio $\frac{p}{2a}$ for the regular polygon as the number of sides increases. What measurement does this ratio approach?

16. Write this ratio using C for circumference and d for diameter of the circle.

5

Transformations

Slide, Turn, and Flip: Exploring Transformations

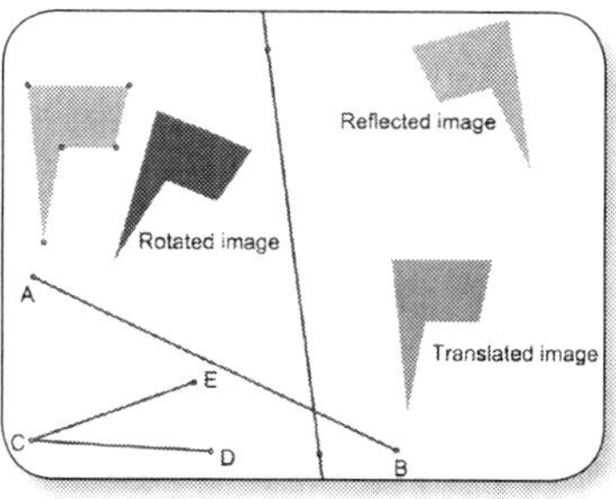

Students explore three basic transformations on a polygon. First they use a vector to translate the polygon, then they use an angle to rotate it, and finally they use a mirror line to reflect it.

Mirror rorriM: Properties of Reflections

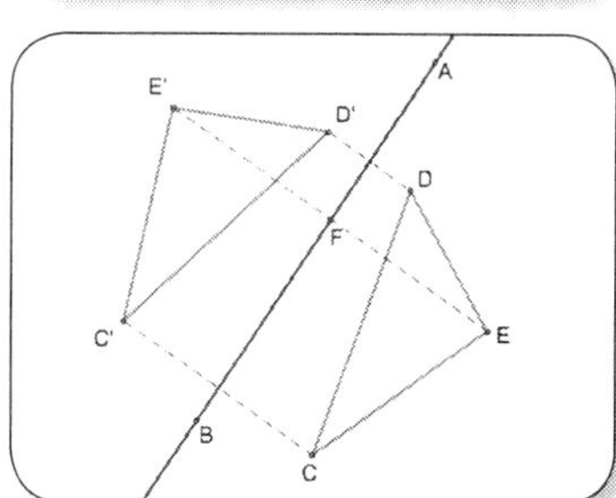

Students explore the properties of reflections with mirror writing and by reflecting a triangle and measuring the sides and angles of the triangle and its reflection. They explore the segments connecting the triangle's vertices and the reflections of the vertices.

Number Flips: Reflections in the Coordinate Plane

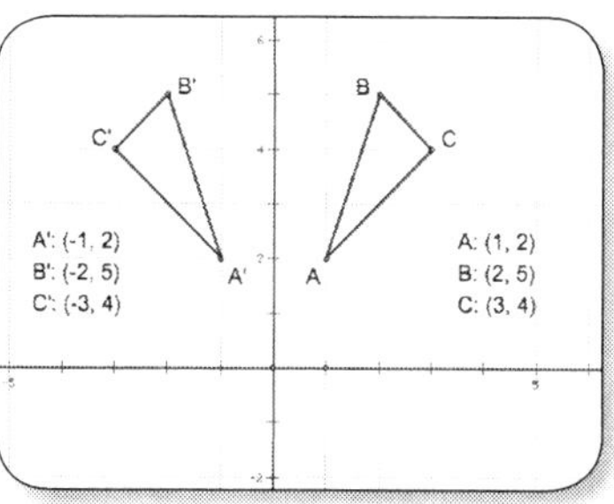

Students construct a triangle in the coordinate plane and measure its coordinates. They reflect it across the *x*-axis and then the *y*-axis, measure the new coordinates, and look for relationships between the coordinates.

Number Slides: Translations in the Coordinate Plane

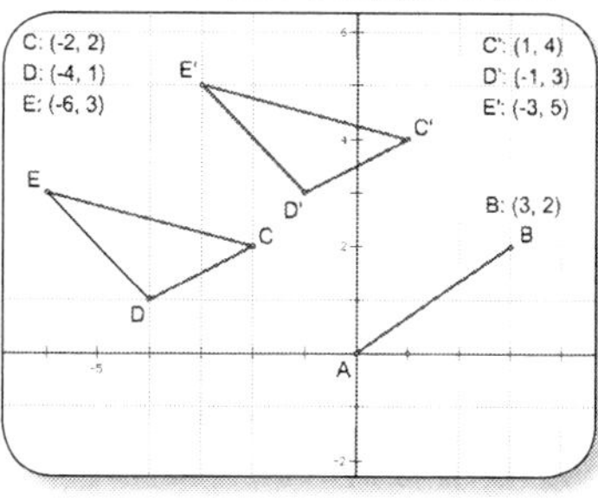

Students construct a segment from the origin to another point in a coordinate plane and mark it as a vector. They construct a triangle, translate it by the vector, and look for relationships among the coordinates of the original vertices, those of the translated image, and the vector.

Floor Tiles: Tessellating with Regular Polygons

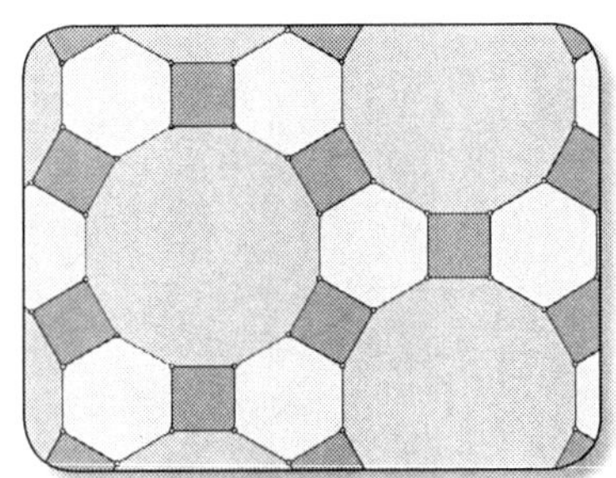

Using custom tools to construct regular polygons, students explore tessellations based on regular polygons. They determine which ones can be used to tessellate a plane and explore why some regular polygons can and others cannot.

Transformers: Exploring Coordinate Transformations

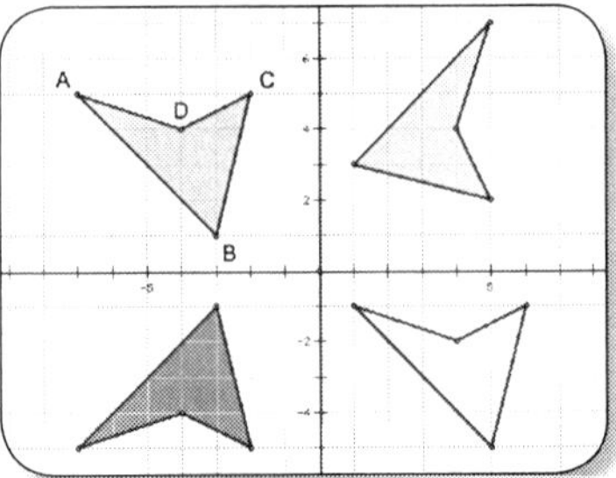

Students use "transformers," which are custom tools that reflect, translate, or rotate a given shape. They measure the coordinates of the shapes produced by the transformers and compare them with the coordinates of the original shapes.

POLYGON polygon: Introducing Dilation

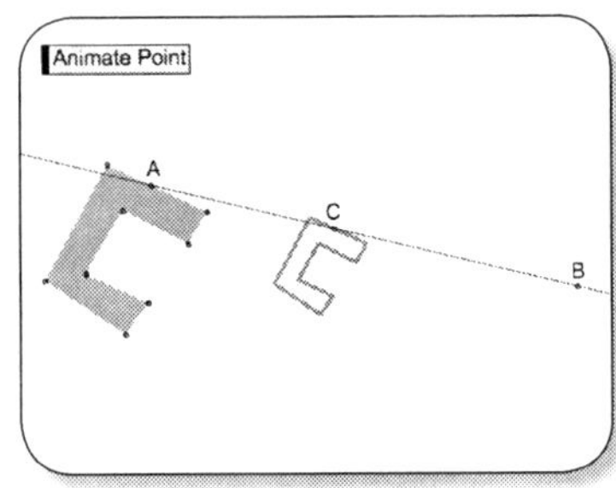

Students explore the properties of dilation by making a dilated image of a polygon. By dragging a point on the image closer or farther from the center of dilation, students realize that the dilated image is similar to its pre-image and that each point of the image is on the line connecting its corresponding point on the pre-image with the center of dilation.

Twist and Shrink: Rotations and Dilations

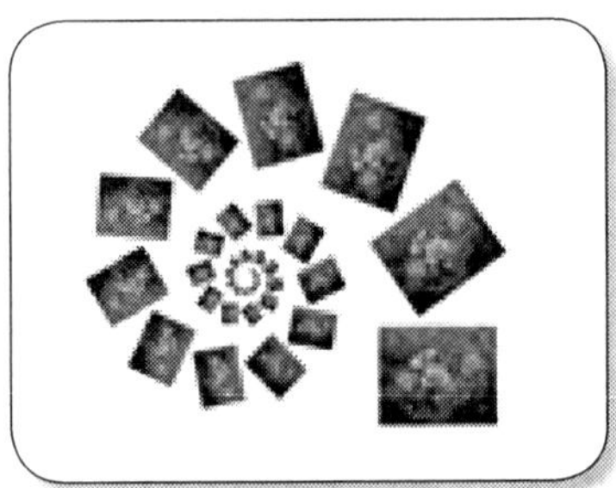

Students create and apply a two-step transformation that combines a rotation with a dilation. They add this new transformation to the Transform menu and then apply it repeatedly to a picture to obtain an eye-catching spiral with interesting mathematical properties.

Menagerie: Comparing Transformations

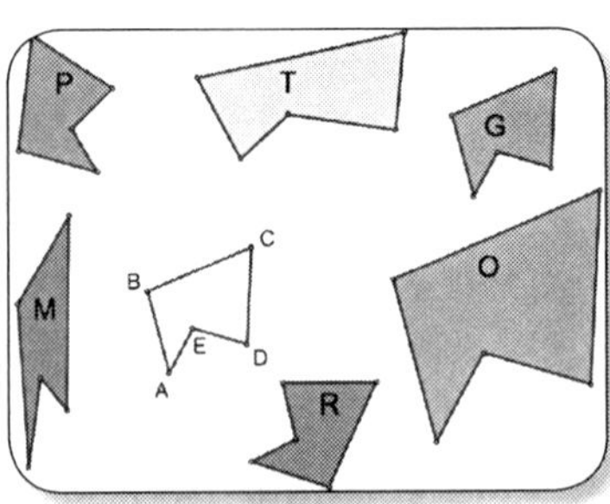

Students explore a collection of shapes that illustrate various geometric transformations of a pentagon. As students drag the vertices of the pentagon, they view the simultaneous changes to the other shapes. Students observe which transformations preserve a polygon's size and shape (translation, rotation, and reflection) and which preserve only a polygon's shape (dilation).

Slide, Turn, and Flip: Exploring Transformations

INTRODUCE

Project the sketch for viewing by the class. Expect to spend about 5 minutes.

These three transformations are called *isometries* because the shape or size of the figure does not change during the transformation. They are also known as *congruence transformations.*

1. Open Sketchpad and enlarge the document window so it fills most of the screen. As you demonstrate, make lines thick and labels large for visibility.
2. Explain, ***Today you are going to use Sketchpad to explore three types of transformations. A transformation is a change in the size or position of a shape. Three basic transformations that maintain the size and the shape of a figure are a slide, also called a* translation; *a turn, also called a* rotation; *and a flip, also called a* reflection.** Write the definition for *transformation* on chart paper.
3. ***Before you start the activity, I will demonstrate how to construct the polygon to be transformed.*** Model the polygon construction in worksheet steps 1 and 2. Here are some tips.
   - In worksheet step 1, explain that students will make a flag-shaped polygon. ***You will use this shape because it is easy to keep track of where it is pointing.*** Hold down the Shift key as you construct the points; they will stay selected as you construct new ones. If you do this, be sure to construct the points in order around the shape. Otherwise you won't be able to construct the interior correctly in worksheet step 2.
   - In worksheet step 2, construct the interior of the polygon. Tell students that they must select the five vertices, in order, clockwise or counterclockwise around the shape.
   - ***After a transformation, this original flag-shaped polygon is called the* pre-image *and the new polygon is called the* image.**
4. ***As you explore the three types of transformations today, predict what will happen before you transform the flag-shaped polygon each time. Then after each transformation, compare the image and the pre-image.***
5. If you want students to save their work, demonstrate choosing **File | Save As,** and let them know how to name and where to save their files.

DEVELOP

Expect students at computers to spend about 25 minutes.

6. Assign students to computers. Distribute the worksheet. Tell students to work through step 32 and do the Explore More if they have time. Encourage students to ask their neighbors for help if they are having difficulty with Sketchpad.
7. Let pairs work at their own pace. As you circulate, here are some things to notice.
   - In worksheet step 7, have students select a new color for the image. It will be easier to observe changes in the transformation if the pre-image and image are different colors. (This holds true for worksheet steps 17 and 26 as well.)
   - In worksheet step 8, tell students that they can use the **Text** tool to drag the image's label off the shape, if needed.
   - In worksheet step 9, encourage students to spend time dragging point *B* up and down as well as left and right. ***Drag to change the size and the direction of $\overline{AB}$, the vector.*** By observing the different locations, students will begin to understand how the image slides in the same direction and distance as the marked vector.
   - In worksheet step 12, encourage students to drag the points to test that their angle construction holds. Students can use **Edit | Undo** if they make a mistake.
   - In worksheet step 14, the center of rotation is not on the original polygon. This is more of a challenge for students to visualize. Students often think that the center must be on the object being rotated. By presenting it this way, this misconception is avoided.
   - In worksheet step 19, have students initially drag point *D* so that $\angle ECD$ is a familiar angle, such as a 90°, 180°, or 45° angle. It may be easier to visualize what happens to the image. Also, be sure students drag point *D* so that the length of $\overline{CD}$ changes but the measure of $\angle ECD$ does not change. ***What happens to the image?***
   - In worksheet step 25, after the reflection, the image may end up off the screen. If that happens, tell students that they can use the scroll bar to see the rest of the image, or have them move the pre-image closer to the mirror line.

- In worksheet step 28, be sure students drag the mirror line in different ways. ***What happens when the mirror line is a horizontal line? When it is a diagonal? When it passes through the pre-image?*** It is important that students realize that lines of reflection are not just vertical or horizontal.
- If students have time for the Explore More, encourage them to use different shapes in their designs and observe what happens during each transformation.

8. If students will save their work, remind them where to save it now.

SUMMARIZE

Project the sketch. Expect to spend about 15 minutes.

9. Gather the class. Students should have their worksheets with them. Open **Slide Turn Flip Present.gsp** and go to page "Translate." Begin the discussion by pressing *Translate.* Change the vector, press *Reset,* and then press *Translate* again. Do this several times with different directions and lengths of the vector. ***What do you notice about the relationship between the vector and the translation?*** Discuss how the image moves the same distance and in the same direction as the vector. Explain that each translation has a length and a distance. ***By marking the segment as a vector, we tell Sketchpad how far and in what direction to translate the shape.***

10. ***What did you notice about the image and the pre-image in a translation?*** Make a chart, such as the following one, completing it as students volunteer how the image and pre-image are the same and how they are different.

Translation

Same	Different
size shape direction they face	location

11. Next, go to page "Rotate" and press *Rotate.* Change the angle of rotation, press *Reset,* and then press *Rotate* again. Do this several times with different angles of rotation. ***What do you notice about the relationship between the angle measure and how far the image rotates?*** Students should observe that the image rotates the same degree as the

angle measure. ***By marking the angle as an angle of rotation, we tell Sketchpad how many degrees to rotate the shape.***

12. Now move the center of rotation, press *Reset*, and then press *Rotate* again. ***What does it mean to be the center of rotation?*** Discuss how it is the point about which the shape is rotated. Reiterate that every rotation has a center of rotation as well as an angle of rotation.

In formal mathematical terms, a rotation is an orientation-preserving transformation, whereas a reflection is not. If you rotate your right hand, it is still a right hand, but if you reflect it, it becomes a left hand. In general usage, however, the *orientation* of an object refers to the position or direction in which it lies, in which case a rotated image would have a different orientation than its pre-image.

13. ***What did you notice about the image and the pre-image in a rotation?*** Make a chart, such as the following one, completing it as students volunteer how the image and pre-image are the same and how they are different. Students might say, *The rotated image is turned,* or *The rotated image is at a different angle,* or *The rotated image is pointing in a different direction.* Although they probably won't say a rotated image has a different orientation, you should be aware of the ambiguity of the term *orientation* (see margin note).

Rotation

Same	Different
size shape	image is turned direction they face location

14. Next, go to page "Reflect" and press *Reflect.* Change the location and direction of the mirror line, press *Reset,* and then press *Reflect* again. Do this several times with different mirror lines. ***What do you notice about the relationship between the mirror line and the image?*** Discuss how the image is flipped over the mirror line the same distance as the pre-image. Each point on the image is the same distance from the mirror line as its matching point on the pre-image. ***By marking the line as a mirror line, we tell Sketchpad how to reflect, or flip, the pre-image.*** Students may understand this concept more easily if they think about a "mirror" image or a "folded" image. Recap by explaining that all reflections have a mirror line, or a line of reflection.

15. ***What did you notice about the image and the pre-image in a reflection?*** Make a chart, such as the following one, completing it as students volunteer how the image and pre-image are the same and how they are different.

Reflection

Same	Different
size shape	image is flipped direction they face (mirror images) location

16. Now go to page "Transformations" and review worksheet steps 31 and 32. Use the buttons to look at one transformed image at a time. ***Is it possible for the translated image to lie on top of the pre-image?*** [Yes, when the translation vector has length zero] ***Is it possible for the rotated image to lie on top of the pre-image?*** [Yes, when the angle of rotation is 0° or 360°] ***Is it possible for the reflected image to lie on top of the pre-image?*** [No] Then show all three transformed images simultaneously. Students should find that the rotated and translated images lie on top of one another when they lie on top of the pre-image.
17. If time permits, discuss the Explore More. ***Tell me about the transformations you used in your design.***
18. ***How would you describe a translation, a rotation, and a reflection to someone? In your description, explain how they are similar and how they are different.*** You may wish to have students respond individually in writing to this prompt.

EXTEND

What questions occurred to you about transformations? Encourage curiosity. Here are some sample student queries.

What happens when you combine different transformations?

What happens if you perform the same transformation twice?

How do different transformations affect different shapes?

ANSWERS

10. The pre-image and the image have the same shape and size, and they both point in the same direction. The only difference between them is that they are in different locations. The image is a copy of the pre-image after a slide.

20. The pre-image and the image have the same shape and size, but point in different directions. The image is a copy of the pre-image after it is turned.

29. The pre-image and the image have the same shape and size, but the image is a mirror image. The image is a copy of the pre-image after a flip.

31. The rotated image and the reflected image have the same shape and size, but the reflected image is a mirror image of the rotated image. The reflected image has been flipped, whereas the rotated image has been turned.

32. The translated image lies directly on top of the pre-image when the translation vector has length zero. The rotated image lies directly on top of the pre-image when the angle of rotation is 0° or 360°. The reflected image cannot lie directly on top of the pre-image unless the pre-image has reflection symmetry and the mirror line is a line of symmetry.

33. Answers will vary. Ask students to describe the different transformations in their designs.

Slide, Turn, and Flip

Name:

A *transformation* is a way of moving or changing a figure. There are three types of basic transformations that preserve the size and shape of the figure: a slide, or *translation;* a turn, or *rotation;* and a flip, or *reflection.* In this activity you'll explore these three transformations using a flag-shaped polygon.

CONSTRUCT

First, construct the flag-shaped polygon. You'll construct the vertices of the polygon and then its interior.

1. Construct five vertices to form a flag shape like the one shown in the picture at right.

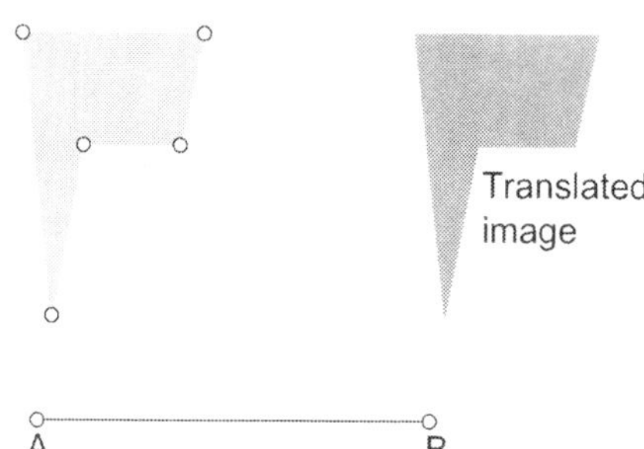

2. Construct the interior of the polygon by selecting all five vertices in order and choosing **Construct | Pentagon Interior.**

Now investigate a slide, or a translation. In order to translate a shape, you'll need to indicate a direction and a distance.

3. To do this, construct $\overline{AB}$.

4. Label the endpoints *A* and *B* by clicking them.

5. Select, in order, point *A* and point *B* and choose **Transform | Mark Vector.** A brief animation indicates that you've marked a vector. Nothing else will happen in your sketch yet.

6. Select the interior of the polygon, and then choose **Transform | Translate.** In the Translate dialog box, make sure **Marked** is checked as the Translation Vector, and then click **Translate.**

7. Now you will change the color of the translated image. With the interior selected, choose **Display | Color** and select a different color.

8. Label the image *Translated image* by double-clicking the interior of the polygon and editing its label.

EXPLORE

9. Drag point *B* to change your vector. Observe the relationship between the original figure, called the pre-image, and the translated image.

10. Compare the translated image to the pre-image. How are they different? How are they the same?

11. Select $\overline{AB}$, including both endpoints, and the translated image, and choose **Display | Hide Objects.** (Do not delete them because you will need them again later.)

CONSTRUCT

Now you'll explore a turn, or a rotation. In order to rotate a shape, you need to indicate a center of rotation and an angle of rotation.

12. Create $\angle ECD$ using two attached segments.

13. Label the vertices of the angle as shown in the picture.

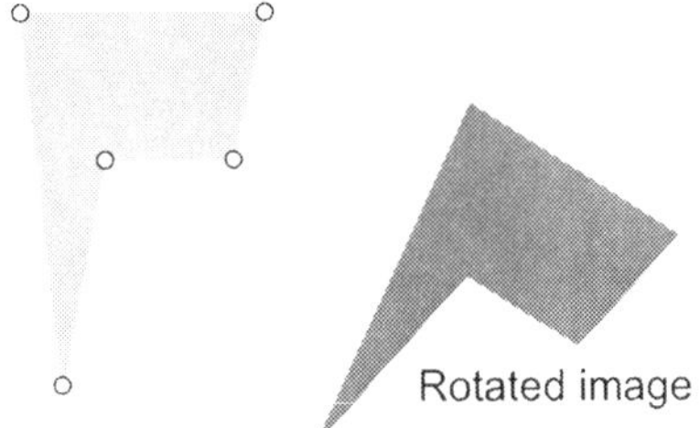

14. Double-click point *C* to mark it as a center of rotation. The point will flash briefly to indicate that it is marked.

15. Now you'll mark $\angle ECD$ as an angle of rotation. Select, in order, points *E*, *C*, and *D*. Then choose **Transform | Mark Angle.**

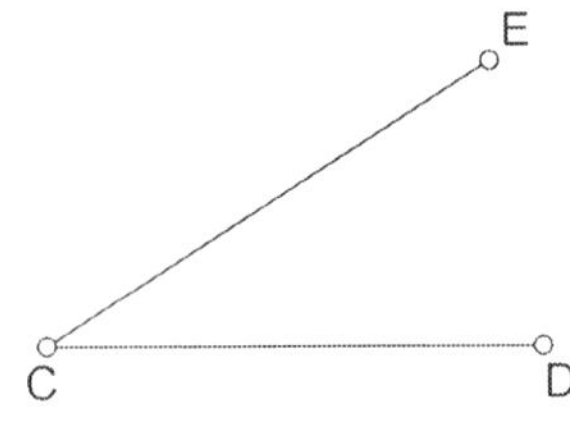

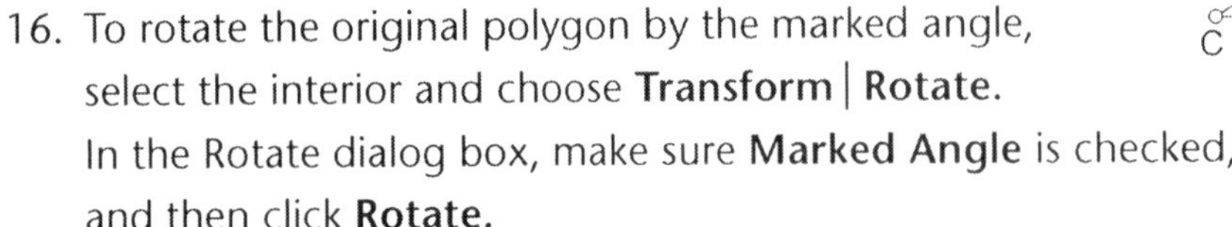

16. To rotate the original polygon by the marked angle, select the interior and choose **Transform | Rotate.** In the Rotate dialog box, make sure **Marked Angle** is checked, and then click **Rotate.**

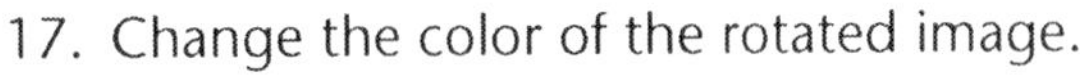

17. Change the color of the rotated image.

18. Label the image *Rotated image.*

EXPLORE

19. Drag point *D* to change your angle. Observe the relationship between the rotated image and the pre-image.

20. Compare the rotated image to the pre-image. How are they different? How are they the same?

21. Hide $\angle ECD$, including all three points, and the translated image.

CONSTRUCT

Now you'll investigate a flip, or a reflection. To reflect a shape, you need a *mirror line* (also called a *line of reflection*).

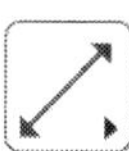

22. Construct a vertical line.

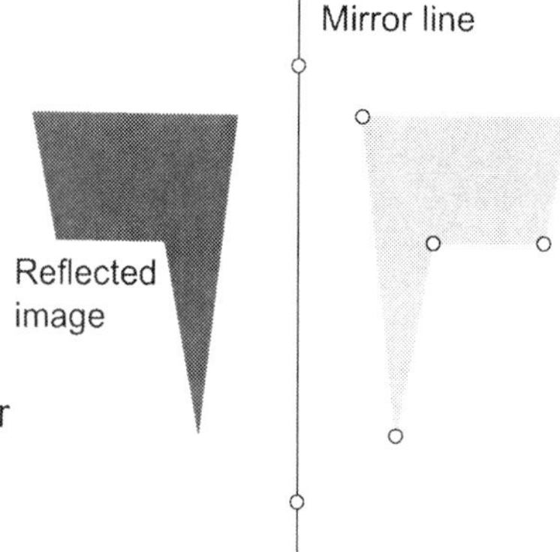

23. Label the line *Mirror line.*

24. Double-click the line to mark it as a mirror. The line will flash briefly to indicate that it is marked.

25. Now you'll reflect the original polygon. Select the interior and choose **Transform | Reflect.**

26. Change the color of the reflected image.

27. Label the image *Reflected image.*

EXPLORE

28. Drag your mirror line. Observe the relationship between the reflected image and the pre-image.

29. Compare the reflected image to the pre-image. How are they different? How are they the same?

Now you'll compare the three transformed images to each other and to the pre-image.

30. Choose **Display | Show All Hidden** to unhide the translated and rotated images.

31. Compare the reflected image to the rotated image. How are they different? How are they the same?

32. Explain whether it is possible for any of the three transformed images to lie directly on top of one another or on top of the pre-image. Experiment by dragging different parts of your sketch.

EXPLORE MORE

33. Use translations, rotations, reflections, or combinations of these transformations to make a design. Use at least one example of each transformation in your design. Be prepared to identify each transformation.

INTRODUCE

Project the sketch for viewing by the class. Expect to spend about 5 minutes.

1. Open Sketchpad and enlarge the document window so it fills most of the screen. As you demonstrate, make lines thick and labels large for visibility.

2. Explain, ***Today you are going to use Sketchpad to explore the properties of reflections. When you look at yourself in a mirror, how far away does your image in the mirror appear to be? Why is it that your reflection looks just like you, but backward? Reflections in geometry have some of the same properties of reflections you observe in a mirror. Before you begin, I will demonstrate how to construct a point and its reflection.*** Model the construction in worksheet steps 1–5. Here are some tips.

   - In worksheet step 4, tell students that another name for a mirror line is a *line of reflection.* Write a definition for *line of reflection* on chart paper. Here is a sample definition: A mirror line, or a line of reflection, is a line that a figure is flipped across to create a mirror image, or a reflection, of the original figure.

   - In worksheet step 5, prior to reflecting point *C*, ask students to predict where the image will appear. Ask a volunteer to construct a point on the sketch to represent the possible image. ***How did you decide where to construct your point?*** The student may respond with the following comment: *I just imagined where the point would be if I folded a piece of paper along the mirror line. The points would match up.* Now select point *C* and choose **Transform | Reflect.** Note how close the student's point was to the image, but let students spend time exploring before talking about any properties of reflections.

   - Drag point *C* and the mirror line so students can see that *C′* also moves.

   - If needed, tell students that in the reflected image, the corresponding point to point *C* is point *C′* and is read "*C* prime." Also explain that after a reflection, the original shape is called the *pre-image* and the reflected shape is called the *image.*

3. ***As you use Sketchpad to explore reflections, think about what conjectures you can make about the properties of reflections.***

4. If you want students to save their work, demonstrate choosing **File | Save As,** and let them know how to name and where to save their files.

DEVELOP

Expect students at computers to spend about 25 minutes.

5. Assign students to computers. Distribute the worksheet. Tell students to work through step 28 and do the Explore More if they have time. Encourage students to ask their neighbors for help if they are having difficulty with the construction.

6. Let pairs work at their own pace. As you circulate, here are some things to notice.

    - In worksheet step 1, students set the Sketchpad Preferences to label points automatically as the points are constructed. Remind students that they can choose the **Text** tool and double-click a label if they need to edit it.

    - In worksheet step 7, when students drag point *C* to trace their names, the image may not appear within the document window. If that happens, tell students to move the mirror line to the center of the sketch *before* tracing.

    - In worksheet step 12, have students drag the vertices of the triangle to test that the construction holds.

    - In worksheet step 13, be sure students select all three vertices and all three sides before reflecting the triangle. If students do not select the vertices, just the segments will be reflected. Tell students they can choose **Edit | Undo,** if needed, and try again.

    - In worksheet step 14, encourage students to drag the mirror line in different directions. ***Can you move the mirror line so the image and the pre-image are close together? Far apart? On top of one another? What happens when the mirror line is a horizontal line? When it is a diagonal? When it passes through the pre-image?*** It is important that students realize that lines of reflection are not just vertical or horizontal and that the distance between the pre-image and the mirror line dictates the distance between the image and the mirror line.

    - In worksheet step 15, tell students that they can use the **Arrow** tool to drag corresponding measurements next to each other. It may be easier to compare the measurements this way.

- In worksheet step 16, review the term *corresponding angle* if students have trouble identifying the angles. ***Look at angle EDC. What is the matching, or corresponding, angle in the image triangle?*** [Angle $E'D'C'$]
- In worksheet step 17, be sure students drag the pre-image and its image to change side lengths and angle measures.
- In worksheet step 20, students will discover that the orientation of the image is opposite that of the pre-image.
- In worksheet step 27, have students try dragging different parts of the sketch to change the measure of $\angle AFE$. Then ask students what the measure of $\angle AFE'$ is. ***If you know the measure of angle AFE, can you figure out the measure of angle AFE'?*** [90°] ***What does this tell you about the mirror line?*** Based on the angle measures, students should realize that the mirror line is perpendicular to a segment connecting a point and its reflected image.
- ***How is the distance from point E to the mirror line related to the distance from the mirror line to point E'?*** [They are the same distance.]
- If students have time for the Explore More, review the definition of an isosceles triangle. ***What is an isosceles triangle?*** [A triangle with two sides of equal length] ***What do you know about the base angles?*** [They have equal measure.]

7. If students will save their work, remind them where to save it now.

SUMMARIZE

Project the sketch. Expect to spend about 15 minutes.

8. Gather the class. Students should have their worksheets with them. Begin the discussion by opening **Mirror rorriM Present.gsp** and use it to support the class discussion.
9. Drag different parts of the sketch and then discuss worksheet steps 18 and 19. ***What effect does reflection have on lengths and angle measures?*** [The lengths and angle measures stay the same.] ***What does this tell you about the image and the pre-image?*** [They are the same size and shape, or congruent.]

10. Now discuss worksheet step 20. ***How are the vertices of the pre-image oriented?*** [Clockwise] ***How are the vertices of the image oriented?*** [Counterclockwise] ***What conjecture can you make about the orientation of the image in relation to the pre-image?*** [The orientation of the image is opposite of the pre-image.]

11. Have students explain their answers to worksheet step 28. ***What did you discover about the mirror line and its relationship to a segment connecting a point and its reflected image? Explain how you know.*** Students may make the following statement: *The mirror line and the segment form right angles, so the mirror line is perpendicular to the segment. The mirror line also cuts the segment in half, because each length from the point of intersection to an endpoint is equal. That makes the mirror line a perpendicular bisector.*

12. ***What conjectures did you make about the properties of reflections?*** Write "In a Reflection, . . ." at the top of a piece of chart paper. Add students' comments. Students may make the following conjectures.

    An image is the same size and shape as its pre-image.

    An image is oriented opposite its pre-image.

    Each point on an image and its corresponding point on its pre-image are the same distance from the mirror line.

    The mirror line forms right angles with a segment drawn between corresponding points on an image and its pre-image.

    The mirror line is a perpendicular bisector of the segment drawn between a point on the image and the corresponding point on the pre-image.

13. If time permits, discuss the Explore More. ***Explain how you constructed your isosceles triangle. What properties of reflections did you use?***

14. ***Describe how a shape is reflected.*** You may wish to have students respond individually in writing to this prompt.

EXTEND

1. ***What questions occurred to you about reflections?*** Encourage curiosity. Here are some sample student queries.

    Do any shapes look the same after they are reflected?

    What happens to a shape if you reflect it across the same line twice?

What happens to a shape if you reflect it across two different lines?

What does the reflection look like if the mirror line is the line of symmetry of the pre-image?

2. Have pairs work together and open a new sketch. The first student does not look at the computer while the second student constructs a shape, reflects it across a mirror line, and then hides the mirror line. The first student must then construct the correct mirror line for the image and its pre-image. Students choose **Display | Show All Hidden** to check how accurate the first student's mirror line was. Have students trade places, so each gets a turn trying to construct the correct mirror line.

ANSWERS

8. Point C' traces the mirror image of the student's name.

18. Reflection preserves lengths and angle measures.

19. A figure and its reflected image are always congruent.

20. The vertices of $\triangle CDE$ go from C to D to E in a clockwise direction. The vertices of the reflected $\triangle C'D'E'$ go from C' to D' to E' in a counterclockwise direction.

28. The mirror line is the perpendicular bisector of any segment connecting a point and its reflected image.

29. Reflect a point across a line. Connect the point with its image point. Also connect each of these points with a third point on the line.

Mirror rorriM

Name:

In this activity you'll investigate the properties of reflections that make a reflection the "mirror image" of the original.

CONSTRUCT

1. In a new sketch, choose **Edit | Preferences | Text** and check **Show labels automatically: For all new points.**

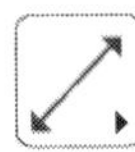

2. Construct vertical line *AB.*

3. Construct point *C* to the right of the line.

4. Mark $\overleftrightarrow{AB}$ as a mirror line by double-clicking on the line. The line will flash briefly to indicate that it is marked.
5. Select point *C* and choose **Transform | Reflect.**
6. Select points *C* and *C′* and choose **Display | Trace Points.** A checkmark in the menu indicates that the command is turned on.

   Choose **Display | Erase Traces** when you wish to erase your traces.

EXPLORE

7. Drag point *C* so that it traces out your name.
8. What does point *C′* trace?
9. For a real challenge, try dragging point *C′* so that point *C* traces out your name.

CONSTRUCT

10. Turn off tracing for points *C* and *C′* by selecting the points and choosing **Display | Trace Points** to remove the checkmark.

11. Choose **Display | Erase Traces.**

12. Construct △*CDE.*

13. Now you'll reflect △*CDE* (sides and vertices) across $\overleftrightarrow{AB}$. Select the entire figure and choose **Transform | Reflect.**

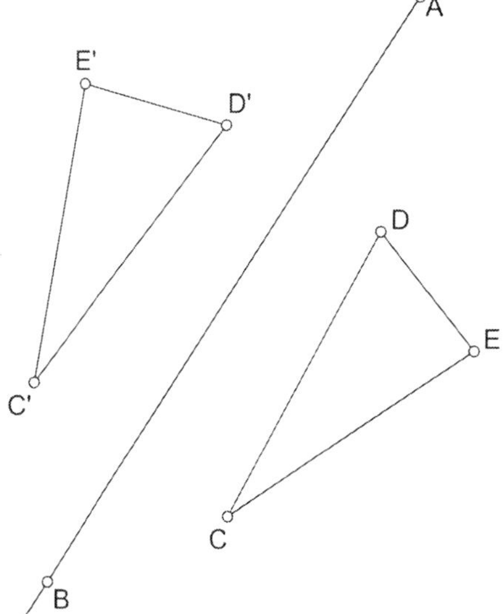

EXPLORE

14. Drag different parts of either triangle and observe how the triangles are related. Also drag the mirror line, $\overleftrightarrow{AB}$.

15. Measure the lengths of the sides of △*CDE* and △*C'D'E'* by selecting each side and choosing **Measure | Length.**

16. Next you'll measure one angle in △*CDE* and measure the corresponding angle in △*C'D'E'*. Select three points that name the angle, with the vertex as your second point, and choose **Measure | Angle.**

17. Drag different parts of either triangle and observe the measurements.

18. What effect does reflection have on lengths and angle measures?

19. Are a figure and its mirror image always congruent? State your answer as a conjecture.

20. Going alphabetically from *C* to *D* to *E* in △*CDE,* are the vertices oriented in a clockwise or counterclockwise direction? In what direction (clockwise or counterclockwise) are vertices *C'*, *D'*, and *E'* oriented?

CONSTRUCT

21. Construct segments connecting each point and its image: *C* to *C′*, *D* to *D′*, and *E* to *E′*.

22. Make these segments dashed by selecting each segment and choosing **Display | Line Style | Dashed.**

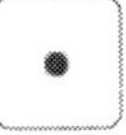

23. Construct point *F* on the intersection of $\overleftrightarrow{E'E}$ and the mirror line.

24. Measure the length of $\overline{E'F'}$ by selecting points *E′* and *F* and choosing **Measure | Distance.**

25. Measure the length of $\overline{EF}$.

26. Measure $\angle AFE$.

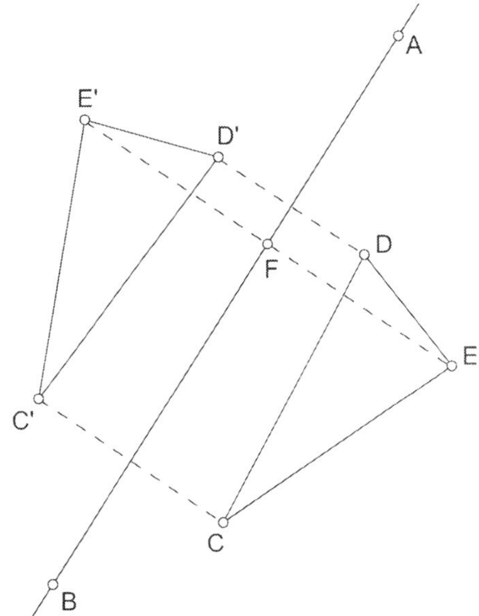

EXPLORE

27. Drag different parts of the sketch around and observe relationships between the dashed segments and the mirror line.

28. How is the mirror line related to a segment connecting a point and its reflected image?

EXPLORE MORE

29. Use a reflection to construct an isosceles triangle. Explain what you did.

Number Flips: Reflections in the Coordinate Plane

INTRODUCE

Project the sketch for viewing by the class. Expect to spend about 5 minutes.

1. Open Sketchpad and enlarge the document window so it fills most of the screen. As you demonstrate, make lines thick and labels large for visibility.
2. Explain, ***Today you're going to use Sketchpad to explore what happens to the coordinates of points when they are reflected across the x- and y-axes in a coordinate plane. I'll model how to construct a shape and reflect it across the y-axis.*** Demonstrate worksheet steps 1–9. Here are some tips.

   - In worksheet steps 1 through 3, explain that that by selecting **Snap Points,** points will only move to where the grid lines intersect, and that by setting the precision of **Others** to **units,** coordinates will appear as integers. After constructing points *A, B,* and *C,* show how when you drag them, they snap to the grid so that their coordinates are always integers.
   - In worksheet step 5, explain that if students need to change a label, they can use the **Text** tool and double-click the label to edit it.
   - In worksheet step 6, review what coordinates are and where the origin is located on the coordinate plane. ***The origin is point* (0, 0). *It is the point where the x- and y-axes intersect.*** Point out where the origin is. ***A point is located using two coordinates. The first coordinate is the x-coordinate and tells how far left or right the point is from the origin. The second coordinate is the y-coordinate and tells how far up or down the point is from the origin.*** Have students tell you the coordinates for the vertex points. Then model how to measure the coordinates of each vertex.
   - In worksheet step 7, say, ***To reflect the triangle, we need to tell Sketchpad what the mirror line, or the line of reflection, will be.*** Double-click the *y*-axis to make it a mirror line. The *y*-axis will flash briefly to indicate that it is marked. Students can also designate a mirror line by selecting it and choosing **Transform | Mark Mirror.**
   - In worksheet step 8, prior to reflecting the triangle, have students predict where the image's vertices will be. ***What do you think the coordinates of the vertices will be in the reflected image?*** You might write down students' predictions on chart paper. Then model how to reflect the triangle across the *y*-axis.

- Label the image's vertices. Review the terminology and notation for reflections. ***The image of point A is point A′; it is read "A prime."***

3. ***Now you'll investigate the relationship between the coordinates of a point and its reflection across both the x- and y-axes. See whether you can make a conjecture based on your findings.***
4. If you want students to save their work, demonstrate choosing **File | Save As,** and let them know how to name and where to save their files.

DEVELOP

Expect students at computers to spend about 20 minutes.

5. Assign students to computers. Distribute the worksheet. Tell students to work through step 15 and do the Explore More if they have time. Encourage students to ask their neighbors for help if they are having difficulty with the construction.
6. Let pairs work at their own pace. As you circulate, here are some things to notice.
   - In worksheet step 8, students will need to select all three vertices and all three segments to reflect the triangle. If they select just the three segments, the points at the vertices will not be reflected. If this happens, have students select the points and reflect them.
   - In worksheet step 11, have students drag the measurements so that corresponding coordinates are next to each other. It may be easier for some students to see the relationship.
   - In worksheet step 12, encourage students to drag the vertices to the four different quadrants on the coordinate plane. Students should observe that the *y*-coordinates stay the same and the *x*-coordinates are opposites.
   - In worksheet step 14, students are asked to predict the coordinates before reflecting the triangle across the *x*-axis. Listen to their thinking; it can help you know whether they understand what is happening to the coordinates during reflection.
   - In worksheet step 15, again have students drag the vertices to the four different quadrants to confirm their predictions or to revise their thinking.

- If students have time for the Explore More, have them think about why the coordinates of the vertices of the image are where they are. ***What is special about the coordinates of the points along the line of reflection?*** [For each point along the line, the *x*- and *y*-coordinates are the same.] Students can check this by measuring the coordinates of points along the line.

7. If students will save their work, remind them where to save it now.

SUMMARIZE

Project the sketch. Expect to spend about 5 minutes.

8. Gather the class. Students should have their worksheets with them. Begin the discussion by opening **Number Flips Present.gsp** and use it to support the class discussion.

9. ***When point C is at* (5, 2), *what do you know about its reflection across the y-axis? Explain your thinking.*** Notice whether students use the term *opposite* rather than *negative.* The opposite of −2 is 2; the opposite of 5 is −5. A negative number is less than zero. Opposite numbers lie the same distance from zero, but on opposite sides, so the numbers have different signs. Here are some sample student responses.

   The reflected point will have the same y-coordinate because it lies along the same horizontal line. All points on the same horizontal line have the same y-coordinate.

   The reflected point will have the opposite x-coordinate because it is the same distance from the y-axis, but in the opposite direction.

10. ***When point C is at* (5, 2), *what do you know about its reflection across the x-axis? Again, explain your thinking.*** Here are sample responses.

    This time C′ will have the same x-coordinate because it is on the same vertical line as point C. Every point on the vertical line has the same x-coordinate.

    The y-coordinates will be opposites. They lie the same distance away from the line of reflection, the x-axis, but on opposite sides.

11. You may wish to have students respond individually in writing to this prompt. ***State a rule for the coordinates of a point P and its reflected image P′ across the y-axis. Now state another rule for the coordinates of a point P and its reflected image, P′, across the x-axis.*** [For any

point $P(a, b)$ that is reflected across the y-axis, the coordinates of P' are $(a, -b)$. For any point $P(a, b)$ that is reflected across the x-axis, the coordinates of P' are $(-a, b)$.]

12. If time permits, discuss the Explore More. ***What did you notice about the coordinates of the vertices of the image and pre-image?*** Students should observe that the x- and y-coordinates are reversed: The x-coordinate in the pre-image becomes the y-coordinate in the image; the y-coordinate in the pre-image becomes the x-coordinate in the image. ***Can you state a rule for the coordinates of any point P and its reflected image, P′, across the line $x = y$?*** [When reflected across the line $x = y$, any point $P(a, b)$ has an image $P'(b, a)$.]

EXTEND

What questions might you ask about reflections in the coordinate plane? Encourage curiosity. Here are some sample student queries.

What happens to the coordinates if you reflect a shape across both the x- and the y-axes?

Can you ever reflect a shape in the coordinate plane so the image and pre-image are the same?

What mirror line would reflect a shape from the first quadrant to the third quadrant? What would happen to the coordinates?

Suppose the mirror line is something like $y = 2$. Can you make a rule for coordinates of any point P and its image P′?

ANSWERS

12. The y-coordinates of the point and its image are the same. The x-coordinates are opposites. A point with coordinates (a, b) has image coordinates $(-a, b)$.

15. This time, the x-coordinates of the point and its image are the same and the y-coordinates are opposites. A point with coordinates (a, b) has image coordinates $(a, -b)$.

17. The coordinates of the image of a point after a reflection across the line $y = x$ are reversed. So a point with coordinates (a, b) has image coordinates (b, a).

Number Flips

Name:

In this activity you'll explore what happens to the coordinates of points when you reflect them across the *x*- and *y*-axes in the coordinate plane.

CONSTRUCT

1. In a new sketch, choose **Graph | Show Grid** and **Graph | Snap Points.**
2. Choose **Edit | Preferences.** On the Units tab in the dialog box, change the precision of **Others** to **units.** Click **OK.** (This will make coordinates appear as integers.)

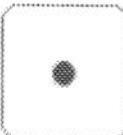

3. Construct points *A, B,* and *C* on the grid.

4. Connect the points with segments to construct $\triangle ABC$.

5. Label the vertices *A, B,* and *C.* Double-click them if you need to change labels.

6. Select each point and choose **Measure | Coordinates.**
7. Double-click the *y*-axis to mark it as a mirror line. It will flash briefly to indicate that it is marked.

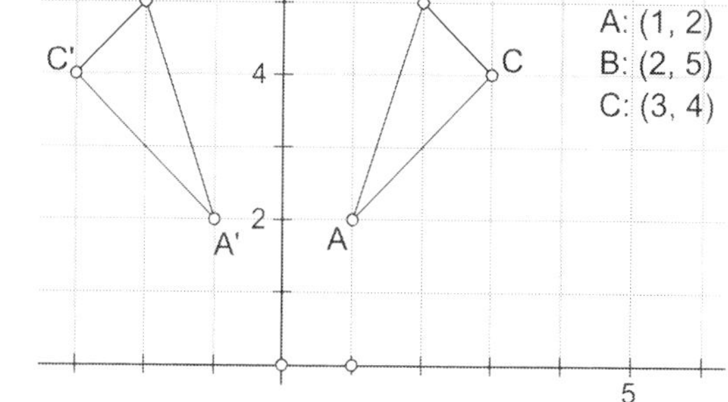

8. Now you'll reflect the triangle across the *y*-axis.

   Select the entire shape and choose **Transform | Reflect.**

9. Label the vertices of the image.

10. Measure the coordinates of the image's vertices.

EXPLORE

11. Drag the vertices and look for a relationship between the coordinates of the original vertices and their reflected images.
12. Describe any relationships between the original coordinates and the coordinates of an image formed by a reflection across the *y*-axis.

13. Now mark the x-axis as a mirror and reflect your original triangle.

14. Before you measure the coordinates of the vertices of the new image, can you guess what they'll be? Measure to confirm.

15. Describe any relationships between the original coordinates and the coordinates of an image formed by a reflection across the x-axis.

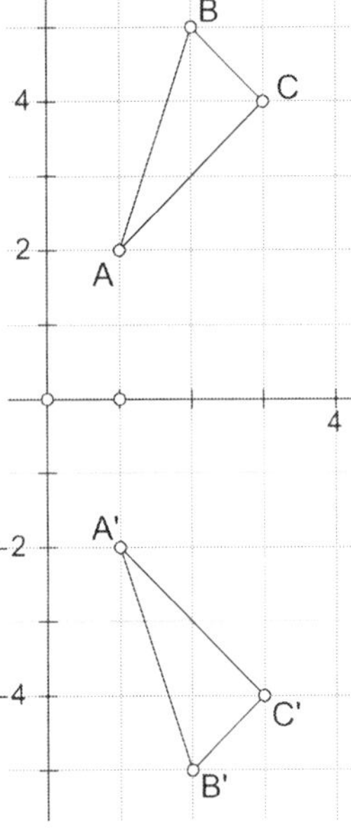

EXPLORE MORE

16. Construct a line on the grid that passes through the origin and makes a 45° angle with the x-axis (in other words, the line $y = x$).

17. Reflect your triangle across this line. What do you notice about the coordinates of the vertices of this image?

Number Slides: Translations in the Coordinate Plane

ACTIVITY NOTES

INTRODUCE

Project the sketch for viewing by the class. Expect to spend about 10 minutes.

1. Open Sketchpad and enlarge the document window so it fills most of the screen. As you demonstrate, make lines thick and labels large for visibility.

2. Explain, ***Today you're going to use Sketchpad to explore a transformation called a slide, or translation, in a coordinate plane. You'll translate a triangle along a vector. You'll explore the relationship among the coordinates of the vector, the pre-image, and its image. Before you begin, I'll demonstrate how to construct the vector and a triangle and how to translate it.*** Demonstrate worksheet steps 1–9. Here are some tips.

   - In worksheet steps 1 and 2, explain that that by selecting **Snap Points,** points will only move to where the grid lines intersect, and that by setting the precision of **Others** to **units,** coordinates will appear as integers.

   - In worksheet step 3, introduce the concept of a vector. ***To translate a shape in Sketchpad, you'll need to construct a segment that you will later mark as a vector. A vector tells Sketchpad how far and in what direction to slide the shape.*** Construct $\overline{AB}$ with point A at the origin and point B anywhere on the grid.

   - In worksheet step 4, model how to label the endpoints. Explain that students can change labels, if needed, by double-clicking on the label and editing it.

You may wish to introduce vector notation. Vector AB is written $\overrightarrow{AB}$.

   - In worksheet step 5, demonstrate how to measure the coordinates of point B. Depending on the background of your students, you might review how coordinates are measured. Drag point B into all four quadrants to show students how the coordinates change.

   - In worksheet step 6, model how to mark $\overline{AB}$ as a vector. ***By selecting first point A and then point B, I am telling Sketchpad that the vector starts at point A and ends at point B.*** Have students look for the brief animation from point A to point B that indicates that the vector has been marked.

   - In worksheet steps 7 and 8, model how to construct $\triangle CDE$ by constructing points on the grid and then segments between them. Students can also select the points and choose **Construct | Segments.**

- In worksheet step 10, model how to select the entire triangle by using the **Arrow** tool to enclose it in a selection rectangle. Before translating the shape, have students predict what will happen. Encourage students to comment on similarities as well as differences between the image and its pre-image. Record students' thoughts on chart paper. Here are some sample predictions.

  The triangle will stay the same size and shape.

  The triangle will be oriented in the same direction. It won't turn or flip.

  The triangle will slide in the direction of the vector to a new location.

  The triangle will slide the same distance as the vector.

  The coordinates of the vertices will change.

  Then demonstrate the translation and check students' predictions.

- Using the **Text** tool, click each vertex of the translated triangle. Review the notation and terms, if needed. ***The vertices are C′, D′, and E′, which are read "C prime," "D prime," and "E prime." The original triangle is called the*** **pre-image, *and the translated triangle is called the*** **image.**

3. ***Now you'll explore on your own. Make some conjectures about the relationship of the coordinates in the image, the pre-image, and the vector.***

4. If you want students to save their work, demonstrate choosing **File | Save As,** and let them know how to name and where to save their files.

DEVELOP

Expect students at computers to spend about 25 minutes.

5. Assign students to computers. Distribute the worksheet. Tell students to work through step 17 and do the Explore More if they have time. Encourage students to ask their neighbors for help if they are having difficulty with the construction.

6. Let pairs work at their own pace. As you circulate, here are some things to notice.

   - Have students enlarge their Sketchpad window as much as possible in order to view the translations better. If a translation lies beyond the document window, students can use the horizontal and vertical scroll bars to see it.

- In worksheet step 6, be sure students select point *A* first and then point *B* when marking the vector. Stress that the direction of the vector is important. ***If you move from point A to point B, in what direction are you moving? Is this the same direction as when you move from point B to point A?***
- In worksheet step 13, tell students that they can drag the corresponding measurements next to one another in order to see the relationships better. Encourage students to drag the points into different quadrants. Also, have them drag point *B* so that it lies on the origin. When this happens, the image and its pre-image will lie on top of one another, and the vector becomes a point, having no direction or distance.
- In worksheet steps 14 and 15, students explore horizontal and vertical translations. When point *B* travels along the *x*-axis, a horizontal translation, the *y*-coordinates stay the same. When point *B* travels along the *y*-axis, a vertical translation, the *x*-coordinates stay the same.
- In worksheet step 16, remind students of the vector's purpose. ***What is the vector's role in a translation?*** [It tells how far and in what direction to translate a shape.] ***If the image is translated left and up, what is the position of the vector?*** [The vector is pointing left and up, so point *B* must be in the second quadrant.]
- For worksheet step 17, students need to understand that the coordinates of the vector tell how far and in what direction to translate each point in the original shape. If the vector starts at the origin (0, 0) and ends at (2, 3), then each point in the triangle will move $2 - 0 = 2$ units in the horizontal direction and $3 - 0 = 3$ units in the vertical direction. ***How can you figure out the horizontal and vertical change from point A to point B?***
- If students have time for the Explore More, they will explore translations with a vector that does not have a starting point on the origin.

7. If students will save their work, remind them where to save it now.

SUMMARIZE

Project the sketch. Expect to spend about 10 minutes.

8. Gather the class. Students should have their worksheets with them. Begin the discussion by opening **Number Slides Present.gsp** and using it to review worksheet steps 13–17. ***How can you use the coordinates of the vector to help you figure out where the triangle will be translated?*** Here are some sample student replies.

   The number of units you go left or right and up or down from point A to point B tells you how many units to translate the triangle.

   Because point A lies on the origin (0, 0), it's easy to see how far to move the triangle. Look at the coordinates of point B. The x-coordinate tells how far to move the triangle horizontally, and the y-coordinate tells how far to move the triangle vertically.

   For any point on the triangle, just add the x-coordinate of point B to its x-coordinate. And do the same for the y-coordinates.

9. If time permits, discuss the Explore More. Have students summarize how to translate the triangle if the starting point of the vector does not lie on the origin. Here is one possible response: *If you find the range of the x-coordinates and the range of y-coordinates of the endpoints of the vector, it will tell you how many units horizontally and vertically to move each point of the triangle.*

10. ***How do you translate points and shapes in a coordinate plane using a marked vector?*** You may wish to have students respond individually in writing to this prompt.

EXTEND

Have students construct a second vector that starts at the ending point of vector *AB* (point *B*) and ends anywhere on the coordinate grid. Students should translate the triangle twice, once along each vector. Ask students to drag point *B*. ***Does dragging point B affect the location of the second translation?*** [No] ***What would a vector look like that would translate the triangle in one translation and give the same outcome as the two translations?*** Students should notice that the two translations combined are equivalent to a single translation by a vector from the origin to the second vector tip.

ANSWERS

14. If you drag point B along the x-axis, the y-coordinates of a vertex and its image point will be the same, but the x-coordinates will differ.

15. If you drag point B along the y-axis, the x-coordinates of a vertex and its image point will be the same, but the y-coordinates will differ.

16. If the vector translates the image up and to the left, point B is in the second quadrant. Its x-coordinate is negative (causing the movement to the left) and its y-coordinate is positive (causing the movement up).

17. The image of (x, y) translated by vector $\langle a, b\rangle$ is the point $(x + a, y + b)$.

18. Find the horizontal and vertical change of the vector by subtracting the coordinates of the starting point from the coordinates of the ending point. Add these values to the coordinates of a point on the pre-image to find the corresponding point on the image. For a horizontal and vertical change of $\langle a, b\rangle$, the coordinates of P' are $(x + a, y + b)$.

Number Slides

Name:

In this activity you'll explore what happens to the coordinates of points when they are translated in the coordinate plane.

CONSTRUCT

1. In a new sketch, choose **Graph | Show Grid** and **Graph | Snap Points.**

2. Choose **Edit | Preferences.** On the Units tab in the dialog box, change the precision of **Others** to **units.** Click **OK.** (This will make coordinates appear as integers.)

3. Construct a segment from the origin to anywhere on the grid.

4. Label the origin *A* and the other endpoint *B.*

5. Measure the coordinates of point *B* by selecting the point and choosing **Measure | Coordinates.**

6. Now you'll mark $\overline{AB}$ as a vector.

   Select point *A* and point *B,* in that order.

   Choose **Transform | Mark Vector.** The segment will animate briefly to indicate that it is marked.

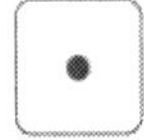

7. Construct points *C, D,* and *E* on the grid.

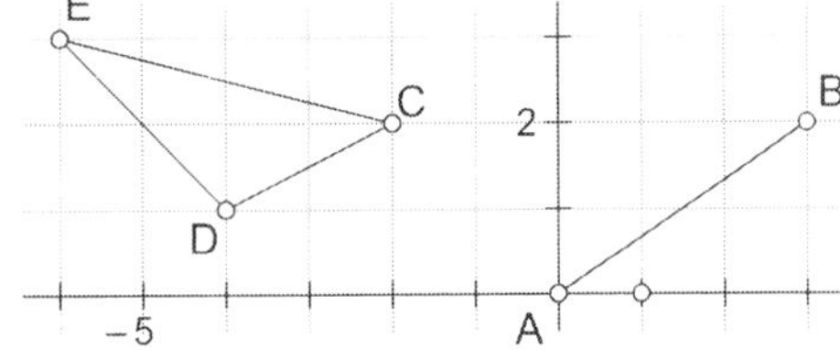

8. Connect the points with segments to construct $\triangle CDE$.

9. Label the vertices *C, D,* and *E.* Double-click them if you need to change labels.

10. Now you will translate the triangle by the marked vector.

    Select the entire shape and then choose **Transform | Translate.**

    In the Translate dialog box, be sure the Translation Vector is set as **Marked.** Then click **Translate.**

11. Label the vertices of the translated image.

12. Measure the coordinates of the three vertices of both triangles.

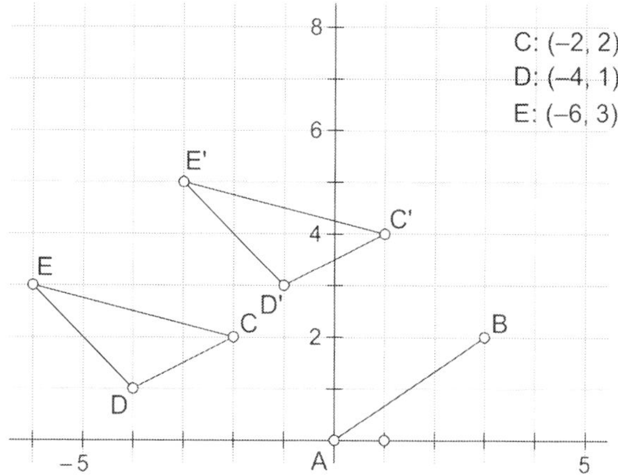

EXPLORE

13. Experiment by dragging point *B* or any of the triangle vertices. Look for a relationship between a point's coordinates and the coordinates of its translated image.

14. Where can you drag point *B* so that the pre-image points and the corresponding image points always have the same *y*-coordinates but have different *x*-coordinates?

15. Where can you drag point *B* so that the pre-image points and the corresponding image points always have the same *x*-coordinates but have different *y*-coordinates?

16. When the vector defined by the origin and point *B* translates the image to the left and up, what must be true of the coordinates of point *B*?

17. Suppose point *A* is the origin and point *B* has coordinates (*a*, *b*). What are the coordinates of the image of a point (*x*, *y*) translated by vector *AB*.

EXPLORE MORE

18. Open a new page by choosing **File | Document Options.** In the dialog box, choose **Blank Page** from the Add Page drop-down menu.

    Now repeat this activity, but in step 3, construct a new segment so neither endpoint lies on the origin. Look for a relationship between a point's coordinates and the coordinates of its image under a translation. Explain how you can find the coordinates of a point P' in the image given point $P(x, y)$ in the pre-image.

Floor Tiles: Tessellating with Regular Polygons

INTRODUCE

Project the sketch for viewing by the class. Expect to spend about 15 minutes.

1. Open **Floor Tiles.gsp** and go to page "Tessellations." Enlarge the document window so it fills most of the screen.
2. Ask, ***What shapes make good tiles for tiling a floor of a bathroom, kitchen, or public building? Why?*** You might also mention the floor of your classroom or hallways if they are tiled. Make sure students understand that a good shape will allow the tiles to fit together without gaps or overlaps. Explain that these tiling patterns are called *tessellations.* If your students are unfamiliar with tessellations, you might show some examples, such as those of Dutch artist M. C. Escher.
3. Explain, ***Today you're going to use Sketchpad to explore which regular polygons tessellate. What do you already know about tessellations?*** Write students' replies on chart paper. Here are some possible student responses.

   It's a pattern formed by shapes.

   There are no spaces between the shapes.

   None of the shapes overlap.

   You can see tessellations everywhere. They are used in tiling floors, making quilts, and laying bricks.

   They are in nature too. Honeycombs, for example, are tessellations.

   You can make a tessellation by flipping, sliding, or rotating a shape.
4. ***Let's write a definition for* tessellation.** Work with students to come up with a definition. Write it in on chart paper. Here is a sample definition: *A tessellation is a pattern of shapes that can cover a region in all directions without any gaps or overlaps.*

The sum of the measures of the interior angles of a polygon can be found using $180(n - 2)$, where n is the number of sides of the polygon.

5. ***A tessellation formed by congruent regular polygons is called a regular tessellation. What is a regular polygon?*** Remind students that a regular polygon has sides of equal length and interior angles of equal measure. If needed, review that *congruent* means "identical in size and shape."
6. ***What is a three-sided regular polygon called?*** [An equilateral triangle] ***What is the measure of each interior angle? How do you know?*** Students should explain that the interior angle measures of a triangle add to 180° and that each angle must measure 60° because the measures are all equal.

7. Start a table on chart paper, such as the following one, and have students fill in the missing information, shown in parentheses in this table.

Regular Polygon	Number of Sides	Interior Angle Measure
Equilateral triangle	3	60°
(Square)	4	(90°)
(Regular pentagon)	5	(108°)
(Regular hexagon)	6	(120°)
(Regular octagon)	8	(135°)
(Regular dodecagon)	12	(150°)

You may wish to have students make predictions about which regular polygons will tessellate before beginning the activity.

8. ***Today you'll explore which of these five regular polygons will tessellate. You'll use custom tools to construct the regular polygons. Before you begin, I will demonstrate how to find and use the custom tool to construct equilateral triangles.*** Model how to construct two attached equilateral triangles in worksheet steps 1–4. As you demonstrate, make lines thick and labels large for visibility. Here are some tips.

   - In worksheet step 2, model how to press and hold the **Custom** tool icon to display the Custom Tools menu. ***Choose* Equilateral Triangle *from the Custom Tools menu.***
   - In worksheet step 3, demonstrate how to make two attached equilateral triangles using the **Equilateral Triangle** tool. Show how the vertex point is highlighted when the pointer is over it. Have students notice in which direction the triangle is constructed.
   - In worksheet step 4, model the drag test. ***Test your construction by dragging several points to be sure the triangles stay attached.*** Explain that if their triangles become unattached, students can use **Edit | Undo** to undo the most recently performed action.

9. If you want students to save their work, demonstrate choosing **File | Save As,** and let them know how to name and where to save their files.

DEVELOP

Expect students at computers to spend about 20 minutes.

10. Assign students to computers and tell them where to locate **Floor Tiles.gsp.** Distribute the worksheet. Tell students to work through step 9 and do the Explore More if they have time. Encourage students to ask their neighbors for help if they are having difficulty with the construction.

11. Let pairs work at their own pace. As you circulate, here are some things to notice.

    - In worksheet step 5, be sure students construct enough triangles to surround two points completely. Have them look at the illustration on their worksheet page for an example. Tell students that the point where the triangles meet is called a *vertex* of the tessellation.
    - In worksheet step 6, ask students to explain how they know an equilateral triangle can tessellate. Students should point out on the sketch that the equilateral triangles do not overlap nor are there any gaps between them.
    - If students have difficulty understanding why equilateral triangles tessellate, ask some pointed questions to help them move their focus from angles within the triangles to angles around the vertices of the tessellation. ***What is the measure of one interior angle of the equilateral triangle?*** [60°] ***How many angles meet at each point, or at a vertex, of the tessellation?*** [Six angles] ***What is the sum of the angle measures around each vertex of the tessellation?*** [360°]
    - In worksheet step 8, students may easily find that the equilateral triangle, the square, and the regular hexagon tessellate, but they may be unsure why. Ask students to look at the relationship between the sum of the angle measures around the vertex of the tessellation and the measure of an interior angle of the regular polygon. Thinking about why some regular polygons don't tessellate can be very insightful. ***Why doesn't a regular pentagon tessellate? What is the sum of the interior angle measures of a pentagon?*** [540°] ***What is the measure of one interior angle of a regular pentagon?*** [108°] ***Two interior angles?*** [216°] ***Three interior angles?*** [324°] ***Four interior angles?*** [432°] ***What does the sum of the angle measures need to be at the vertex of a tessellation?*** [360°]

- In worksheet step 9, students will need to find the measure of one interior angle of the regular heptagon. They can find the sum of the angle measures using $180(7 - 2) = 900$, or 900°. Then they can find each interior angle measure: $900 \div 7 = 128\frac{4}{7}$, or $128\frac{4}{7}°$.
- If students have time for the Explore More, they will construct a regular tessellation by transforming an equilateral triangle. This is a good activity for students who have had experience with translations, rotations, and reflections. ***How can you construct another copy of the equilateral triangle using what you know about translations, rotations, and reflections?***

12. If students will save their work, remind them where to save it now.

SUMMARIZE

Project the sketch. Expect to spend about 10 minutes.

13. Gather the class. Students should have their worksheets with them. ***Which of the regular polygons will tessellate?*** Have volunteers come up and construct tessellations for the equilateral triangle, the square, and the regular hexagon.
14. ***Why did some regular polygons tessellate while others did not? What did you discover?*** Students may make the following observations.

    We noticed that an interior angle measure of the equilateral triangle is 60 degrees, which is a factor of 360. An interior angle measure of the square is 90 degrees, which is a factor of 360. An interior angle measure of the regular hexagon is 120 degrees, which is also a factor of 360. The regular pentagon and regular octagon had interior angle measures that were not factors of 360.

    The angles around a point in the tessellation need to add up to 360 degrees. That means an interior angle measure of a regular polygon must be able to divide into 360 evenly so there aren't any gaps or overlaps.

15. Refer back to the table on chart paper you made earlier. Add a fourth column to the table with the heading "Ratio of Angle Sum at Vertex to Interior Angle Measure." Have students help fill in the column.

Regular Polygon	Number of Sides	Interior Angle Measure	Ratio of Angle Sum at Vertex to Interior Angle Measure
Equilateral triangle	3	60°	360° ÷ 60° = 6
Square	4	90°	360° ÷ 90° = 4
Regular pentagon	5	108°	360° ÷ 108° = 3.33...
Regular hexagon	6	120°	360° ÷ 120° = 3
Regular octagon	8	135°	360° ÷ 135° = 2.66...
Regular dodecagon	12	150°	360° ÷ 150° = 2.4

16. Be sure students grasp the relationships represented in the fourth column of the table. ***How many hexagons meet at a vertex in the tessellation?*** [3] ***What is the measure of one of those angles?*** [120°] ***What is the sum of the angle measures at the vertex?*** [360°] ***Why can't a regular heptagon tessellate?*** [An interior angle measure does not divide evenly into 360°, so you cannot place the shapes together without gaps or overlaps.]

17. ***Could any regular polygon with more than six sides tessellate the plane?*** Help students see from angle measures that there are only three regular tessellations: the equilateral triangle, the square, and the regular hexagon. In the Explore More, students can look for the eight semiregular tessellations (in which the same combination of two or more different types of regular polygons meet in the same order at each vertex), as well as tessellations that have more than one type of vertex arrangement.

18. If time permits, discuss the Explore More. Have students describe how they transformed the equilateral triangle to tessellate it. Students may have translated it, reflected the triangle across a side, rotated it around a vertex or a midpoint of a side, or combined transformations. You may wish to have students come to the projector and model their tessellation constructions.

EXTEND

Do you think every triangle will tessellate? How about every quadrilateral? Use what you have learned to explain your answer. Students should reason that because the angles of a triangle add up to 180°, any triangle can be positioned so each angle appears twice around a vertex so that the sum is equal to 360°. The angles of a quadrilateral add up to 360°, so you can position each angle once at a vertex to equal 360°. Every triangle and every quadrilateral will tessellate.

You can extend this even further to ask whether every hexagon will tessellate, or whether *any* pentagon will tessellate.

ANSWERS

6. Equilateral triangles tessellate because each of their angles measures 60° and their sides are congruent. Because 60° divides evenly into 360°, the angles of the triangles collect around a single point without gaps or overlaps. Because the sides are congruent, they can match up.

8. The only regular polygons that tessellate are those whose angles divide evenly into 360°: equilateral triangles, squares, and regular hexagons.

9. A regular heptagon has angles of $128\frac{4}{7}°$. This means two heptagons could meet to make an angle of $257\frac{1}{7}°$, which is less than 360° and would leave a gap. Three would meet to make an angle of $385\frac{5}{7}°$, which is more than 360° and would make an overlap. So regular heptagons cannot tessellate.

10. Starting with a single equilateral triangle, students can tessellate by reflecting triangles over sides, by translating, by rotating 180° about midpoints of sides, or by rotating by 60° around the triangle's vertices.

11. Answers will vary. There are many possible arrangements. Two or more regular polygons will fit around a vertex if the sum of the interior angles is 360°.

Floor Tiles

Name:

Squares make good floor tiles because they can cover a surface without any gaps or overlaps. This kind of tiling is called a *tessellation*. Are there other shapes that would make good tiles? In this activity you'll investigate which regular polygons tessellate.

CONSTRUCT

1. Open the sketch **Floor Tiles.gsp.** Go to page "Tessellations."

2. Choose the **Equilateral Triangle** tool in the Custom Tools menu. Click twice in the sketch. Pay attention to the direction in which the equilateral triangle is created.

3. Use the **Equilateral Triangle** tool again to construct a second equilateral triangle attached to the vertices of the first triangle.

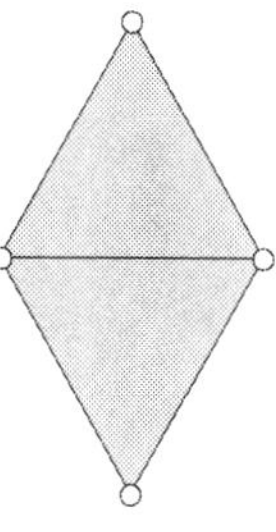

4. Drag several points on the triangles to make sure they are attached. If your triangles don't stay attached, choose **Edit | Undo** until the second triangle goes away and then try again. Always start and end your dragging with the cursor positioned on an existing point.

5. Keep attaching triangles to sides of existing triangles until you have triangles completely surrounding at least two points.

6. So far, you've demonstrated that equilateral triangles can tessellate. You can tile the plane with them without gaps or overlaps. Why do equilateral triangles tessellate? (One explanation involves angle measures.)

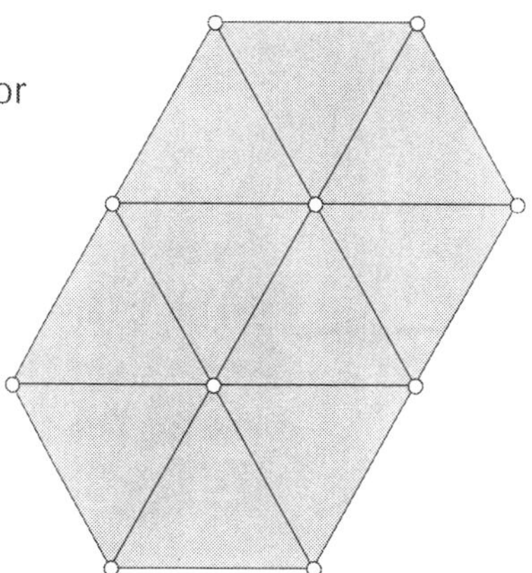

EXPLORE

7. Repeat the investigation with squares, regular pentagons, regular hexagons, regular octagons, and regular dodecagons. Click on the **Custom** tool icon to choose the appropriate custom tool.

8. Which of the regular polygons you tried will tessellate and which won't? Why?

9. Do you think a regular heptagon (seven sides) would tessellate? Explain.

EXPLORE MORE

10. Now see whether you can tessellate an equilateral triangle without using the custom tool. Construct an equilateral triangle (or use the **Equilateral Triangle** tool just once). Then use commands from the Transform menu to create a tessellation. Explain what you did.

11. You can also tile a floor using combinations of regular polygons. Experiment with tessellations that use two or more different regular polygons. How many different types of tessellations can you find? How can you tell whether two or more regular polygons will fit around a vertex?

Transformers: Exploring Coordinate Transformations

INTRODUCE

Project the sketch for viewing by the class. Expect to spend about 5 minutes.

1. Open **Transformers.gsp** and go to page "Across." Enlarge the document window so it fills most of the screen.
2. Explain, ***You have learned about the different properties related to rotation, reflection, and translation. You might even find it easy to tell transformations apart just by looking at shapes and their images. But some applications of transformations, such as video game graphics, require comparing or creating transformations of shapes that you can't actually see. Coordinate geometry provides a very powerful tool to do this.***
3. Drag the vertices of *ABCD* and say, ***Here's a quadrilateral that I can move and change by dragging its vertices. Now I'm going to apply a mysterious transformation to it by using one of my Transformers.*** Model how to use **Transformer #1** by choosing it from the Custom Tools menu and clicking the vertices of *ABCD* in order. ***Which transformation produced this image of ABCD?*** [Reflection across the *x*-axis] Then show students how the labels of the two quadrilaterals can help them figure out which are the corresponding coordinates.
4. Select point *A* and model how to measure its coordinates. Then select point *A1* and ask, ***Can you predict the relationship between the coordinates of point A and point A1? This is the type of relationship you'll be looking for today. Understanding how the coordinates of the original shape are related to those of its image is very helpful because it allows you to transform a shape without finding a line of reflection or a center of rotation.***

DEVELOP

Expect students at computers to spend about 30 minutes.

5. Assign students to computers. Tell them where to locate **Transformers.gsp.** Distribute the worksheet. Tell students to work through step 9. Encourage students to ask their neighbors for help if they are having difficulty with the custom tools.
6. Let pairs work at their own pace. As you circulate, here are some things to notice.
   - Help students use the custom tools correctly by clicking on the vertices of *ABCD in order* around the shape.

- Make sure students choose the **Arrow** tool and drag the vertices of *ABCD* after using a custom tool.
- If students have difficulty seeing the relationship in worksheet step 4, suggest that that they keep *ABCD* in Quadrant II.
- Help students organize the coordinate measures so that they can more easily compare corresponding coordinates.
- Show students that the custom tools can be used on any quadrilateral, not just *ABCD*.

7. When most students have reached worksheet step 9, gather the class together and make sure that they have been able to describe the coordinate relationships in worksheet steps 5 and 8. If students have not done so, you might introduce the notation $(x, y) \to (new\ x, new\ y)$ to describe these relationships. Explain, ***You have looked at two types of reflection that changed either the x-coordinate or the y-coordinate. Can you predict what might happen with other transformations?*** Then ask students to work through step 20 and do the Explore More if they have time.

SUMMARIZE

Project the sketch. Expect to spend about 10 minutes.

8. Ask students to describe how each transformer works. Help them describe the transformations in terms of arbitrary coordinates. For example, **Transformer #1** takes a point (x, y) and reflects it to a point $(x, -y)$. **Transformer #2** takes a point (x, y) to $(-x, y)$. Mention that these descriptions apply only to very specific transformations (reflections over the axes and rotations around the origin), and that reflections across other lines, and rotations around other points, may not be so easily described by coordinates.

9. To summarize the transformations, go to page "Match." Challenge students to describe the Transformers (and transformations) that would transform the quadrilateral to each of the three images. Students may want you to measure the coordinates of the three holes. One of the transformations is an x-shift of 3 and a y-shift of -4; another is a rotation by 270° counterclockwise (or 90° clockwise); and the third is a reflection across the line $y = x$ (which would be easier for students to see if you first assigned Extension 1). Many students may suggest that this last reflection is a rotation by 180°. Use the relationships between

the coordinates they discovered to help them understand why it is not a rotation.

10. If time permits, discuss the Explore More. Invite students to share their observations about sequences of transformations.

EXTEND

1. Have students construct a line that passes through the origin with a slope of 1 (with equation $y = x$). Tell them to mark this line as a mirror and then use **Transform | Reflect** to reflect *ABCD* across this line. Ask them to explore the relationships between the corresponding coordinates. Later, change the slope of the line so that students can see that this relationship will break down.

2. ***What other questions might you ask about transformations?*** Encourage all curiosity. Here are some ideas students might suggest.

   Will it always be true that using two transformations in a row is the same as a single transformation?

   What's going on when I choose a Transformer and don't follow the vertices in order?

   What kind of transformation just switches the two coordinates?

ANSWERS

4. The image is a reflection of *ABCD* across the *x*-axis.

5. The *x*-coordinates are the same and the *y*-coordinates have the opposite sign. Using symbols, $(x, y) \rightarrow (x, -y)$.

7. The image is a reflection of *ABCD* across the *y*-axis.

8. The *x*-coordinates have the opposite sign and the *y*-coordinates are the same. Using symbols, $(x, y) \rightarrow (-x, y)$.

9. Use **Transformer #1** on the quadrilateral in Quadrant I or **Transformer #2** on the quadrilateral in Quadrant III.

11. The image is a translation of *ABCD*.

12. The *y*-coordinates are the same and the *x*-coordinates of the image are 8 greater than those of *ABCD*. Using symbols, $(x, y) \rightarrow (x + 8, y)$.

13. To land the image on the target, use an x-shift of 10.0 and a y-shift of -6.0.

14. Answers will vary, but both the x- and y-coordinates should be negative.

16. The image is a rotation of *ABCD* by 90° counterclockwise around the origin.

17. The x-coordinate of each point in *ABCD* becomes the y-coordinate of the corresponding point in the image. The y-coordinate in *ABCD* becomes the opposite of the x-coordinate in the image. Using symbols, $(x, y) \rightarrow (-y, x)$.

19. The image is a rotation of *ABCD* by 180° around the origin. The x-coordinate of each point in *ABCD* becomes the opposite of the x-coordinate in the image. The y-coordinate becomes the opposite of the y-coordinate in the image. Using symbols, $(x, y) \rightarrow (-x, -y)$.

20. Answers will vary. Sample solutions: Use **Transformer #2** on *ABCD*. Use **Transformer #1** or **Transformer #4** on the image in Quadrant IV. Use **Transformer #5** on the image in Quadrant III. Other combinations are possible.

21. Answers will vary. The example given will produce a rotation of *ABCD* by 180° around the origin. This can also be accomplished by using the two Transformers in the reverse order, or by using **Transformer #5.**

Transformers

Name:

In this activity you'll use transformers—tools that transform a shape into a new shape—to explore the relationship between the coordinates of the original and transformed shapes.

EXPLORE

1. Open **Transformers.gsp** and go to page "Across." Drag the vertices of quadrilateral *ABCD* into any shape you like.

2. Select the vertices of *ABCD* and choose **Measure | Coordinates.**

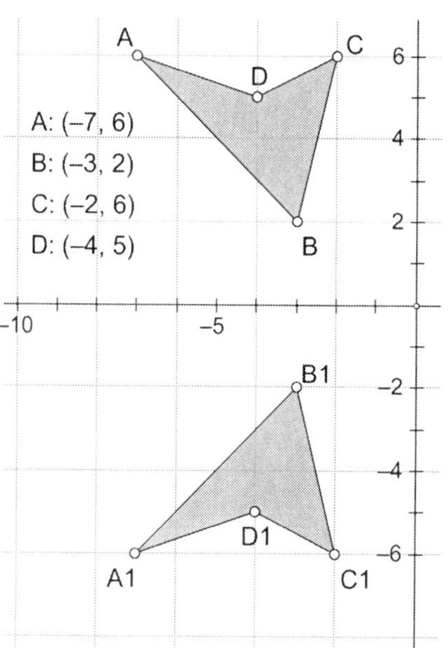

3. Choose **Transformer #1** from the Custom Tools menu. Click each vertex of *ABCD* in a clockwise (or counterclockwise) order. You should see a new quadrilateral, called the *image,* in Quadrant III.

4. Drag the vertices of *ABCD* or the quadrilateral interior. Describe how the two quadrilaterals are related.

5. Measure the coordinates of the image. Describe the relationship between the corresponding coordinates of the two quadrilaterals.

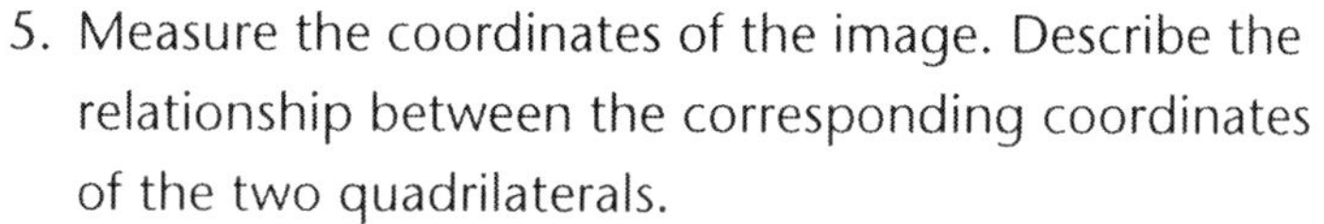

6. Use the custom tool **Transformer #2** on *ABCD.*

7. Drag the vertices of *ABCD.* Describe how the new image is related to *ABCD.*

8. Measure the coordinates of the new image. Describe the relationships between the corresponding coordinates of *ABCD* and the new image.

9. Use either **Transformer #1** or **Transformer #2** to create a quadrilateral in Quadrant IV. Find two different ways of doing this.

Transformers

continued

10. Go to page "Along." Use **Transformer #3** on *ABCD.*

11. Drag the vertices of *ABCD.* Describe how the two quadrilaterals are related.

12. Measure the coordinates of *ABCD* and the image. Describe the relationship between their corresponding coordinates.

13. Experiment with changing the *x*-shift and *y*-shift values. To change a value, either double-click it and enter a new value, or select it and use the + and − keys on your keyboard. Then press *Show Target.* What shift values will make the image land on the target?

14. Drag *ABCD* so that it is in Quadrant I only. What shift values will place the image in Quadrant III?

15. Go to page "Around." Use **Transformer #4** on *ABCD.*

16. Drag the vertices of *ABCD.* Describe how the two quadrilaterals are related.

17. Measure the coordinates of *ABCD* and the image. Describe the relationship between their corresponding coordinates.

18. Use **Transformer #5** on *ABCD.*

19. Measure the coordinates of the new image. Describe how the new image is related to *ABCD* and the relationship between their corresponding coordinates.

20. Find three different ways of producing another image in Quadrant I using any of the Transformers.

EXPLORE MORE

21. Go to page "Explore More." Experiment with different sequences of five Transformers. What happens if you first use **Transformer #1** and then use **Transformer #2** on the image? What other Transformers or sequences of Transformers will accomplish the same result?

INTRODUCE

Project the sketch for viewing by the class. Expect to spend about 5 minutes.

1. Open Sketchpad and enlarge the document window so it fills most of the screen. As you demonstrate, make lines thick and labels large for visibility.
2. Explain, ***Today you are going to use Sketchpad to explore the properties of* dilation. *You'll make a* dilated image *of a polygon, you'll describe how the dilated image is similar to and different from the original, and you'll use your observations to write a definition of* dilation.**
3. If your students are experienced with Sketchpad constructions, explain, ***In steps 1 through 11, you'll construct a polygon, and then use a point on the polygon and a point on a line to make a dilated image of the polygon.***
4. If your students are not experienced with Sketchpad constructions, explain, ***Before you begin, I will demonstrate how to construct a simple shape and dilate it.*** Model worksheet steps 1–12. Here are some tips.
   - In worksheet step 1, construct only three points to make a triangle. Tell students they will construct more points to make a more interesting polygon.
   - After you construct the triangle in worksheet step 2, model dragging a vertex of the triangle to change its shape. Students will need to do this to adjust the shapes (and possibly to uncross the sides) of their polygons.
   - In worksheet step 3, ask, ***How do you think the new point will move?*** Solicit several answers, and then say, ***Let's use a drag test to see.*** Drag the point to see how it behaves. ***Now we'll make the point go around the triangle automatically.***
   - In worksheet step 7, explain, ***We'll use a line to make the dilated image. Here's how to get the* Line tool.** Demonstrate pressing and holding the **Straightedge** tool icon and choosing the **Line** tool.
   - In worksheet step 10, explain, ***Now that one of the line's points follows the triangle, we'll make another point on the line and see how it behaves.***

- In worksheet step 11, ask, ***Can you tell what path the new point is following?*** Solicit several answers without commenting on correctness.
- In worksheet step 12, explain, ***The path point C follows is a dilated image of the path of point A. You'll turn on tracing to see the path more clearly.***

5. ***As you use Sketchpad to explore dilation, think about how you can describe the properties of a dilated image.***
6. If you want students to save their work, demonstrate choosing **File | Save As,** and let them know how to name and where to save their files.

DEVELOP

Expect students at computers to spend about 25 minutes.

7. Assign students to computers. Distribute the worksheet. Tell students to work through step 24 and do the Explore More if they have time. Encourage students to ask their neighbors for help if they are having difficulty with the construction.
8. Let pairs work at their own pace. As you circulate, here are some things to notice.
   - In worksheet step 2, some students will find their polygons are crossed rather than simple polygons. One way they can fix this is by dragging the points so that the sides no longer cross. But if they've already carefully placed the points in a particular shape, they can preserve their work by clicking in empty space to deselect everything and then clicking each point to select them in order before constructing the polygon.
   - In worksheet step 15, point *C* and part of its trace may go beyond the edge of the window. Students can use **Edit | Undo Animate Point** to get *C* back inside the window. To keep the trace within the window, students should first make sure the sketch window is maximized so that it fills the screen. If the trace still goes outside the window, the student can move point *B* closer to *A* (making the image closer to the original polygon), move point *C* closer to *A*, or use the **Arrow** tool to drag the vertices of the original polygon to make it smaller.
   - In worksheet steps 16 and 17, students observe the effects of changing the scale factor for dilation and changing the position of the center

of dilation. Encourage them to pay particular attention to observing carefully and writing precise and detailed answers.

- In worksheet step 18, students observe the effects of a negative scale factor for dilation. Encourage careful observation in this step as well, but don't worry too much if student answers are a bit fuzzy on this question. A precise description and understanding of negative dilation is not crucial for students' understanding at this stage.
- In worksheet steps 22 and 23, students review the effects of changing the scale factor and write a definition of *dilation*. Encourage them in their definitions to distinguish between *dilation* (the process) and *dilated image* (the object that results from dilation).

SUMMARIZE

Project the sketch. Expect to spend about 15 minutes.

9. Gather the class. Students should have their worksheets with them. Open **POLYGON polygon Present.gsp** and use it to support the class discussion.

10. Press *Animate Point* and discuss worksheet step 12. ***How does animating point A affect the position of point C?*** [It moves the line, and *C* has to stay on the line. Some students may observe that when *A* is closer to the fixed point *B*, point *C* also gets a bit closer to *B*, and when *A* is farther from *B*, so is *C*.]

11. Press *Erase Traces* and then press *Show Locus*. Use the locus to discuss worksheet steps 15 and 16. Move *C* so it's on the other side of *A* from *B*. ***What's different about the image when point C is on the other side of A?*** [Instead of shrinking, the image is now stretched so that it's bigger than the original.] ***When is the dilated image smaller than the original?*** [The image is smaller when point *C* is between point *B* and point *A*.] ***When is the dilated image exactly the same size as the original?*** [The image is the same size when point *C* is exactly on top of point *A*.]

12. Use the locus to discuss worksheet step 17. Move point *B* so it's closer to point *A*. ***What's different about the image when point B is close to A?*** [The image gets closer to the original but does not change size.] ***When is the dilated image smaller than the original?*** [The image is smaller when point *C* is between point *B* and point *A*.] ***When is the dilated image exactly the same size as the original?*** [The image is the same size when point *C* is exactly on top of point *A*.]

13. ***What effect does dilation have on lengths and angle measures?*** [The angle measures stay the same, but the lengths change.] ***What do you notice about how the lengths change?*** [They change in a similar way, all getting larger or smaller together.] This is an opportunity to call attention to the use of the word *similar* in describing both how the lengths change in similar ways and how the shape of the dilated image is similar—but not identical—to the shape of the original object. ***What effect does the position of point B have?*** [The dilated image shrinks toward point *B* or stretches away from point *B*.] Tell students that point *B* is called the *center of dilation.*

14. Discuss worksheet step 19. ***What happened when you moved point C to the other side of point B?*** [The image shrank to a point and then flipped and got larger again.] Tell students that this is a very interesting effect, but not one that they need to understand in detail at this stage.

15. ***What observations did you make about the properties of dilation?*** Write "In a dilation . . ." at the top of a piece of chart paper. Add students' comments. Students may make the following conjectures.

    An image is the same shape as its pre-image, but a different size.

    An image is oriented the same way as its pre-image.

    Each point on an image and the corresponding point on its pre-image are on the same line with the center of dilation.

EXTEND

1. ***What questions occurred to you about dilation?*** Encourage curiosity. Here are some sample student queries.

    Do any figures have different shapes after they are dilated?

    What happens to a shape if you dilate it twice, using two different centers?

    Can you use two different dilations so that one makes the dilated image smaller and the second returns the dilated image to the same size as the original?

    When is a flipped dilation the same size as the original?

2. Have pairs work together and open a new sketch. The first student does not look at the computer while the second student constructs a shape, dilates it using a center point, and then hides the center point. The first

student must then construct the correct center point for the image and its pre-image. Students choose **Display | Show All Hidden** to check the accuracy of the first students' center points. Have students trade places, so each gets a turn trying to construct the correct center point.

ANSWERS

12. Point *C* moves all the way around the edges of the polygon.
13. Point *C* leaves a trace behind. The trace looks like an outline of the original polygon, only smaller.
15. With point *C* closer to *B*, the new trace is smaller than the previous trace.
16. With point *C* on the other side of *A*, the new trace is larger than the original polygon.
17. With point *C* closer to *A*, the new trace is still larger, but is closer in size to the original polygon.
18. With point *B* closer to *A*, the new trace is the same size as the last time, but is closer to the original.
19. With point *C* on the other side of *B*, the new trace is flipped.
21. The locus shows the same positions as the traced point did.
23. Students will make a number of observations, including some of these.

    When C is between A and B, the dilated image is smaller than the original.

    When C is on the other side of A from B, the dilated image is larger.

    When C is on the other side of B from A, the dilated image is flipped.

    When C is on top of A, the dilated image is the same size as the original.

    When C is on top of B, the dilated image shrinks to a point.

    The locus always seems to have the same shape as the original polygon, but is different in size. Except when the locus flips, its shape stays right-side up, the same way as the original.
24. Student definitions of *dilation* will vary, but all the definitions should say something about the dilated image being smaller or larger than the original, but having the same shape.

POLYGON polygon: Introducing Dilation

Name:

In geometry, *dilation* is one way of changing a figure. In this activity you'll *dilate* a polygon, observe the result, and write your own definition of *dilation*.

CONSTRUCT

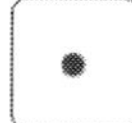

1. In a new sketch, construct seven or eight points in an interesting shape.

2. Select all the points and choose **Construct | Polygon Interior.** If some of the edges cross each other, drag points to untangle the polygon.

3. With the polygon selected, choose **Construct | Point on Polygon Interior.** Drag the point to see how it moves.

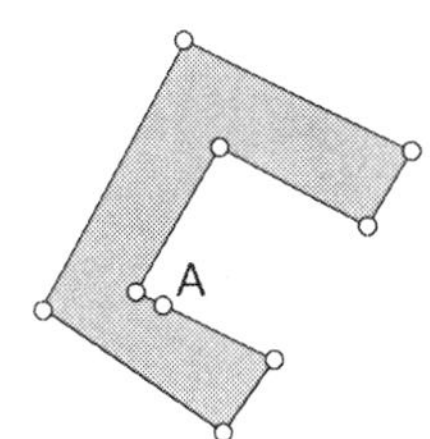

4. Label the new point *A*.

5. Now you'll create an animation button.
   With point *A* still selected, choose **Edit | Action Buttons | Animation.** In the dialog box that appears, click **OK.** A new button appears, labeled *Animate Point.*

6. Press *Animate Point* and observe how point *A* moves. Press *Animate Point* again to stop the motion. Leave *A* on a side of the polygon, not too close to a vertex.

7. Construct a line by clicking on point *A*, dragging, and releasing well off to the side of the polygon.

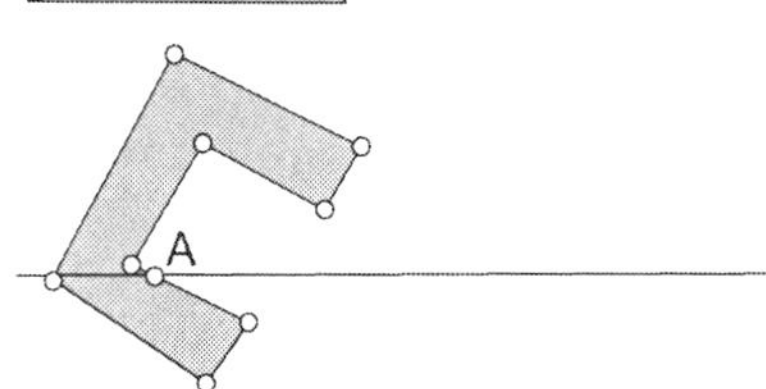

8. Label the second point *B*.

9. Press *Animate Point* and observe how the line behaves as *A* moves around the polygon. Let it go all the way around and then stop the animation.

10. Construct a point on the line about halfway between *A* and *B*.

11. Label this point *C*.

EXPLORE

12. Press *Animate Point* and observe the path of point *C*. How would you describe its motion?

13. The shape point *C* makes is a *dilated image* of the path of point *A*. To see more clearly how *C* behaves, stop the animation, select point *C*, and choose **Display | Trace Point.** Press *Animate Point* again, and describe the result.

14. Move *C* closer to point *B*. Then choose **Display | Erase Traces.**

15. Animate again. How is the new trace different?

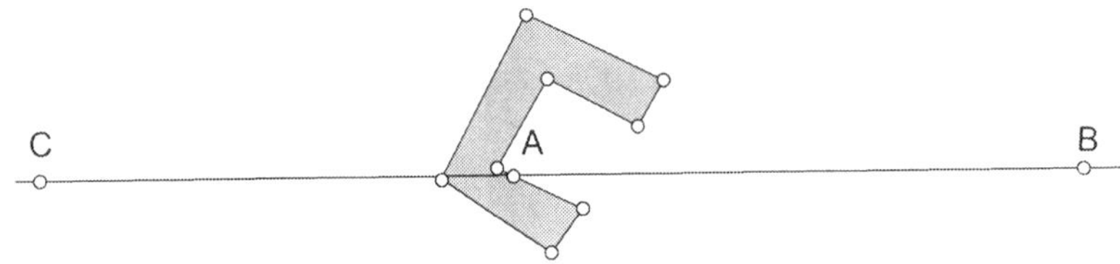

16. Move *C* to the other side of the polygon from point *B*. Erase the traces and animate again. How is this trace different? (You may need to adjust the position of point *B* or the size of your polygon to keep the trace on the screen.)

17. Drag point *C* closer to *A* and animate again. How is this trace different?

18. Drag point *B* closer to *A* and animate again. How is this trace different?

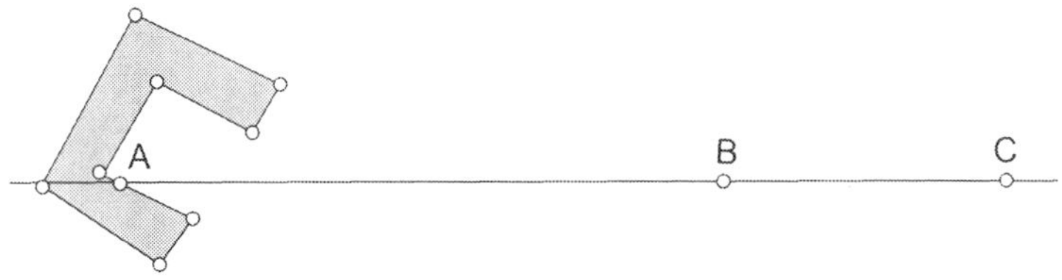

19. Drag *C* past *B*, so that it's on the other side of *B* from the polygon. Erase the traces and animate one more time. How is this trace different?

20. Drag *C* back to a position between *B* and the polygon, and erase the traces.

21. Select both *A* and *C* and choose **Construct | Locus.** Then press *Animate Point.* What relationship is there between the traced point and the locus?

22. Turn off tracing by selecting *C* and choosing **Display | Trace Point.** Then erase the traces.

23. Drag point *C* back and forth along the line. What do you observe about the locus? Drag *C* to many different positions, and write down as many observations as you can about the appearance of the locus and its relationship to the path of point *A*. How is the locus the same as the path of *A*? How is it different?

24. *Dilation* is the geometric term for the relationship between the path of point *A* and the locus. Using your answer to the last question, write your own definition of *dilation.* As part of your definition, explain what a *dilated image* is.

EXPLORE MORE

You can dilate the entire polygon rather than just its outline (the path of point *A*). To do so, follow these steps.

25. Select in order points *B, A,* and *C* and choose **Transform | Mark Ratio.**

26. Select point *B* and choose **Transform | Mark Center.**

27. Select the polygon interior and choose **Transform | Dilate.**

28. Drag point *C* back and forth along the line. How does the dilated image of the polygon compare to the original polygon?

Twist and Shrink: Rotations and Dilations

ACTIVITY NOTES

INTRODUCE

Project the sketch for viewing by the class. Expect to spend about 5 minutes.

1. Open Sketchpad and enlarge the document window so it fills most of the screen.
2. Explain, ***Today you are going to use Sketchpad to create some eye-catching mathematical designs. You'll start with just a single picture. By applying two simple transformations to the picture over and over again, you'll create a design that looks very complex and striking. The mathematics of your design, however, is very understandable.***
3. Review with students the meaning of *rotation* and *dilation*. The term dilation may be new to students. If so, ask them whether they've heard the term used in contexts other than mathematics. Some students may recall that an eye doctor may dilate their eye to make its pupil bigger.

DEVELOP

Expect students at computers to spend about 30 minutes.

4. Assign students to computers. Distribute the worksheet. Tell students to work through step 24 and do the Explore More if they have time. Encourage students to ask their neighbors for help if they are having difficulty with the construction.
5. Let pairs work at their own pace. As you circulate, here are some things to notice.
   - In worksheet step 2, students drag or paste a picture into their sketch. They'll rotate and dilate this picture many times, filling the sketch with transformed images. If students' sketches get crowded, suggest that they make the window larger or the picture smaller.
   - In worksheet step 6, students rotate their picture. If the rotated picture is not visible on the screen, students can drag their picture closer to point B and/or move $\angle ABC$ closer to the center of the sketch window.
   - In worksheet step 7, students explore the concept of rotation. Ask them if the length of the segments that form $\angle ABC$ have any effect on the angle of rotation.
   - In worksheet step 8, students rotate their picture by 180° to turn it upside down. They may expect that if they keep dragging the angle past 180°, the picture will continue to rotate all the way around, but the measurement is always between 0° and 180°.

Angle markers, a Sketchpad feature not covered in this activity, are one way to create angles whose measurements range from 0° to 360°.

- In worksheet step 14, students explore the effects of dilating their picture by a scale factor between 0 and 1. Ask students to drag the point on their segment so that the scale factor is 0.5. ***How do you think the size of the original picture and its dilated image compare?*** [The length and width of the dilated image are half the original dimensions.]
- In worksheet steps 17–18, students rotate a point *G* by the marked angle *ABC* and then dilate the result, point *G'*, by the marked scale factor to obtain *G''*. By performing this two-step transformation on a point, students are able to save their steps as a custom transformation in step 19. Now, when students look in the Transform menu, they'll see **Twist and Shrink** as one of the transformations available to them.
- In worksheet step 22, students create a spiral by repeatedly applying their custom transformation. For best viewing, students may need to make some small adjustments to their sketch. They can make the original picture larger or smaller by selecting it, holding the Shift key, and dragging a corner of the selection frame. They can drag $\angle ABC$ to the center of the sketch window and adjust the angle by dragging point *A* or point *C*. Or they can make the scale factor at least 0.9 so that the complete set of transformed images is visible.
- In worksheet step 23, students experiment with their spiral of transformed images by changing the amount of rotation and the scale factor. There are lots of striking designs that students can make. If students would like to save particular designs, they can take note of the rotation angle and scale factor to recreate the design later. Alternatively, they can choose **File | Document Options** and in the Add Page pop-up menu choose **Duplicate.**
- The Explore More explains how students can swap one image for another without having to rebuild their spiral from scratch. By selecting any image in their spiral and pasting another picture onto it, students can update the entire spiral of images with the new picture.

SUMMARIZE

Project the sketch. Expect to spend about 10 minutes.

6. Gather the class. Ask students to share some of the spirals they built. If none of their models are available, open **Twist and Shrink Example.gsp.**

7. Discuss step 24 of the worksheet. One way to think about what's happening with the spiral of images is to start with the scale factor equal to 1. In this case, the original picture is repeatedly rotated about point *B*, but it doesn't get any closer to the point. When the scale factor is less than 1, each application of "Twist and Shrink" brings it ever closer to point *B*. The smaller the scale factor, the more quickly the transformed images approach point *B*.

8. Discuss step 25 of the worksheet. For convenience, use **Twist and Shrink Example.gsp** to explore students' thinking. To see the images form clear lines coming out from point *B*, it's best to make the scale factor at least 0.9. Angles like 30°, 45°, and 60° (all of which are factors of 360) form lines. One way to think about this is to make the scale factor equal to 1. In this case, the Twist and Shrink custom transformation just rotates the original picture again and again. Factors of 360 will result in the rotated images overlapping each other.

   Interestingly, there are also other angles that produce lines of images. For example, when $\angle ABC$ is 144°, the first four applications of the Twist and Shrink transformation place images at 144°, 288°, 432°, and 576°. The next rotation places an image at 720°, which is equivalent to a rotation of 0° and overlaps the original picture. From here, the rotation cycle starts all over again. In general, rotation angles whose ratio to 360° reduces to a fraction with a small denominator create lines.

EXTEND

Ask students to design their own two-step custom transformations by combining other pairs of transformations together. They might, for example, translate a picture by a marked vector and then reflect it across a line to form a glide reflection.

ANSWERS

7. When point *A* or *C* is dragged directly toward or away from *B*, the measure of $\angle ABC$ stays about the same and the rotated picture stays in place. In practice, it's hard to keep the angle measure exactly the same when dragging.

8. Rotating the picture by 180° turns it upside down.

14. The dilated image changes size, but not shape, as the scale factor changes. When the scale factor approaches 1, the dilated image moves toward the original picture, growing until it is the exact same size as the original picture and overlaps it when the scale factor equals 1. When the scale factor approaches 0, the dilated image moves toward point *B*, growing smaller and smaller until it disappears entirely when the scale factor equals 0.

21. Answers will vary.

24. See step 7 of the Summarize section.

25. See step 8 of the Summarize section.

Twist and Shrink: Rotations and Dilations

Name:

In this activity you'll rotate a picture by a marked angle and dilate a picture by a marked scale factor. Then you'll combine both transformations into a single two-step transformation that rotates a picture and dilates the result. By repeating this process, you'll create a dramatic result.

ROTATE A PICTURE

You'll begin by picking a picture, measuring an angle, and rotating the picture.

1. In a new sketch, choose **Help | Picture Gallery.**
2. Drag a picture into your sketch (or copy it and paste it into your sketch).

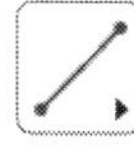

3. Construct two segments that share a common endpoint.

4. Press and drag to form a selection rectangle around the vertex of the angle. Choose **Measure | Angle.**

5. Select point *B* and choose **Transform | Mark Center.**
6. Select your picture and choose **Transform | Rotate.** With the dialog box open, click the angle measurement in the sketch. Then click **Rotate.**
7. Drag points *A* and *C*. What type of motion makes the rotated picture move? What type of motion keeps it nearly steady? Why?
8. How much of a rotation do you need to turn the picture upside down?

DILATE A PICTURE

Now you'll measure a ratio and mark it as a scale factor to dilate your picture.

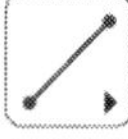

9. Construct a segment.

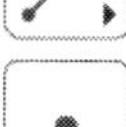

10. Construct a point on the segment.

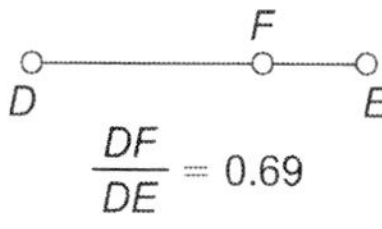

11. With the point selected, choose **Measure | Value of Point.**
12. With the value selected, choose **Transform | Mark Scale Factor.**
13. Select your picture and choose **Transform | Dilate.** Click **Dilate.**

14. Drag the point on your segment. What is the effect on the dilated image? What happens as the point approaches one endpoint or the other?

DEFINE A CUSTOM TRANSFORMATION

Now you'll combine the rotation and the dilation into a single transformation that will allow you to rotate the picture and then dilate the rotated image.

15. Delete all pictures except the original picture.

16. Construct a point and label it *G*.

17. Select point *G* and choose **Transform | Rotate.** Click **Rotate.** Label the rotated point *G*′.
18. Select point *G*′ and choose **Transform | Dilate.** Click **Dilate.** Label the dilated point *G*″.
19. Select points *G* and *G*″ and choose **Transform | Define Custom Transform.** Name your transformation **Twist and Shrink** and click **OK.**
20. Select points *G*, *G*′, and *G*″ and choose **Display | Hide Points.**
21. Select the picture and choose **Transform | Twist and Shrink.** With the newest picture selected, choose **Twist and Shrink** again. If you keep twisting and shrinking, what do you predict the result will look like?

22. Apply **Transform | Twist and Shrink** again and again to create at least 30 transformed pictures.
23. Drag points in your sketch to change the angle and ratio of your custom transformation. Look for angles and ratios that make interesting spirals.
24. Drag point *A* to make $\angle ABC$ smaller and smaller. Explain why the transformed pictures wind their way towards point *B*.
25. What combination of angles and ratios make straight lines of images come out from point *B*? Can you explain why?

EXPLORE MORE

26. Find a new picture and copy it. Then select any picture in your spiral and choose **Edit | Paste Replacement Picture** to create a new spiral.

Menagerie: Comparing Transformations

INTRODUCE

Project the sketch for viewing by the class. Expect to spend about 5 minutes.

1. Open **Menagerie.gsp.** Enlarge the document window so it fills most of the screen.
2. Go to page "Shapes." Explain, ***Here's a pentagon, which I can change by dragging any of its vertices. Today you'll look at a wide variety of transformations of this pentagon ABCDE. Before we start, can anyone think of a particular transformation that you've encountered before?*** Students might mention some of the more common congruence-preserving transformations such as rotation, reflection, and translation. After these three have been listed, ask, ***Does anyone know whether there are other kinds of transformations? If yes, how many do you think there might be?*** Students might think of dilation, or perhaps shearing. ***You're going to use Sketchpad to investigate some new transformations that you may never have seen before. You're going to focus especially on how each of these transformations changes ABCDE in specific ways. Before you begin, I'll demonstrate how the sketch works.***
3. Press *Show Menagerie.* Show students how they can drag the vertices of *ABCDE* to make all the shapes on the screen change. You might want to refer to these shapes using their labels or their unique colors. Press *Hide Menagerie.* Drag *ABCDE* so that it forms a recognizable shape such as a square. ***Can anyone guess what some of the other shapes will look like?*** Give students a chance to sketch out a few suggestions and ask for examples. Tell students, ***This may seem like a difficult task now, but it'll become much easier as you explore these shapes.***

DEVELOP

Expect students at computers to spend about 15 minutes.

4. Assign students to computers and tell them where to locate **Menagerie.gsp.** Distribute the worksheet. Tell students to work through step 8 and do the Explore More if they have time.
5. Let pairs work at their own pace. As you circulate, here are some things to notice.
   - Encourage students to drag *ABCDE* as they work through the worksheet. This will help them develop more general patterns and relationships.

- For worksheet step 3, students might have a difficult time finding any relationships. You can help them by directing their attention to geometric properties such as length, angle, area, and size. It may also help students to drag *ABCDE* into more recognizable shapes.
- For worksheet step 4, at first students may think that *R* is a rotation of *ABCDE*. Encourage them to keep changing *ABCDE* and watch how *R* changes. Use the word *flipped* or describe *R* as facing the other way. These are both good descriptions. It might be helpful also for students to realize that both *G* and *P* can be shifted along the plane to lie on top of *ABCDE,* but that *R* has to lift out of the plane and reflect to get back to *ABCDE.*
- For worksheet steps 5 through 7, encourage students to look for examples in one of the shapes. This will not be possible for worksheet step 6, although students may be able to think of a sheared shape that has the same side lengths as *ABCDE* with different angles. As needed, advise students to measure areas in worksheet step 7.
- For worksheet step 8, students can use a mix of formal and informal words to describe the transformations.

SUMMARIZE

Project the sketch. Expect to spend about 10 minutes.

6. Gather the class. Students should have their worksheets with them. Begin the discussion by reviewing worksheet step 3. Ask for volunteers to describe how the shapes resulting from transformations that don't preserve congruence are related to *ABCDE.*
7. Hide the shapes that are not congruent to *ABCDE* (shapes *M, T,* and *O*). Ask students to describe each visible shape and how it was transformed from *ABCDE.* List the properties each shape has in common with *ABCDE* and the properties it does not share.
8. If your students are familiar with the Transform menu commands, invite them to construct shapes on the projected sketch that result from the same types of transformations as the shapes already in the sketch, but that occupy different locations.
9. If time permits, discuss the Explore More. The vertices of shape *X* are the intersections of lines formed through this complex series of constructions: a line through the pre-image vertex and the center of a circle; the perpendicular to this line through the center; a line through

the pre-image vertex and the intersection of the perpendicular with the circle; a mirror line through the intersections of two lines with the circle; and a reflection of a line across the mirror line. The *Hint* buttons show this sequence for point *A*.

EXTEND

1. Invite students to describe another transformation, different from those in the shapes. You may challenge them to come up with another transformation that preserves congruence (the only other one would be a glide reflection) or other transformations that preserve only, for example, parallel lines (like a sheer transformation).
2. ***What other questions might you ask about transformations?*** Encourage all inquiry. Here are some ideas students might suggest.

   Don't congruent figures look alike? Are reflections really congruent?

   Is there a way, other than eyeballing, to know whether one shape is a transformation of another and what kind of transformation it is?

   How many different kinds of transformations are there?

ANSWERS

2. *R, P, G*
3. Answers may vary. *O* is a dilation of *ABCDE* by a factor of two. *T* has a horizontal dilation but no vertical dilation (it's a stretched version of *ABCDE*). *M* has a horizontal dilation of 0.5, but a vertical dilation of 2 (it's stretched vertically but squished horizontally).
4. *R* has a different orientation from the other shapes. By dragging *ABCDE* into a rectangle or square, *R* would no longer appear to have a different orientation.
5. Yes, *O* has the same angles, but the side lengths are twice as long.
6. No
7. Yes. *M* has the same area as *ABCDE*, but the two shapes are not congruent.
8. *P* is a rotation. *G* is a translation. *R* is a reflection. *O* is a dilation. *T* and *M* are both transformations that do not preserve angle, length, or shape.

Menagerie

Name:

In this activity you'll explore several different transformations of a pentagon.

EXPLORE

1. Open **Menagerie.gsp** and go to page "Shapes." Drag the vertices of pentagon *ABCDE*. Press *Show Menagerie.*

2. Drag the vertices of *ABCDE* again and observe what happens to each of the shapes. List the shapes that are congruent to *ABCDE.*

3. For each of the shapes that you did not list in worksheet step 2, describe the features it has in common with *ABCDE.*

4. Shapes *P* and *G* are congruent to *ABCDE.* All three are congruent to shape *R* too, but *R* has a distinguishing feature. Explain how it's different. How could you change the shape of *ABCDE* so that *R* would not seem to have this distinguishing feature?

5. Do any of these shapes have the same angle measures as *ABCDE,* but different side lengths? If so, which ones?

6. Do any of these shapes have the same side lengths as *ABCDE,* but different angle measures? If so, which ones?

7. Do any of these shapes have the same area as *ABCDE* without being congruent? If so, which ones?

8. Use what you know about each shape to figure out what type of transformation relates it to *ABCDE.*

EXPLORE MORE

9. Go to page "Explore More." This time *ABCDE* has been transformed in quite a different way. Drag the vertices of the pentagon to see how shape *X* changes. What relationships can you find between *X* and *ABCDE*? Look carefully at length, angle, and size. Compare your findings with the shapes you investigated on the previous page.

10. Press *Show Objects.* Describe what new relationships you see.

Three-Dimensional Solids and Volume

Cube Nets: Three-Dimensional Puzzles

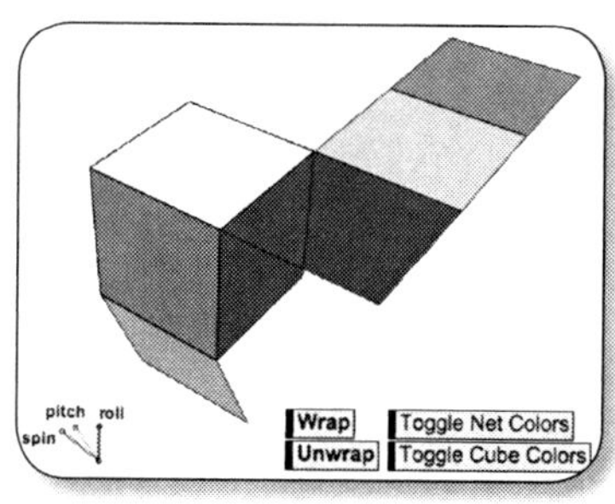

Students are challenged to match the faces of a cube net with the corresponding faces of a cube that can be manipulated dynamically and viewed from any direction. The colored or patterned faces of the given nets and cubes provide just enough information to match each pair of corresponding faces.

Prism Nets: Spatial Visualization

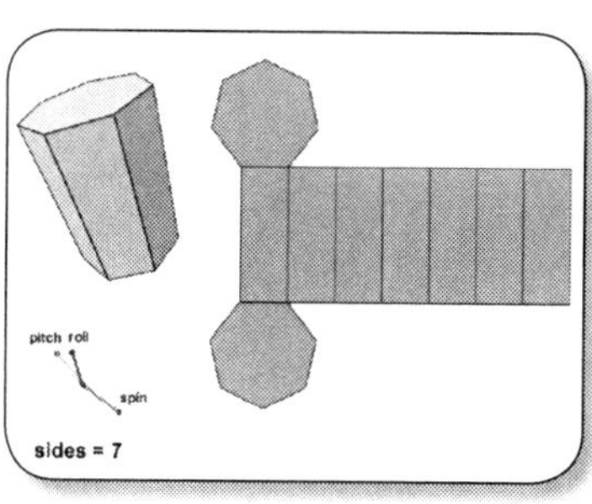

Students relate a three-dimensional view of a regular right prism to the two-dimensional net from which it can be folded. The model allows students to view the prism from different angles and change the height, the size of the base, and the number of sides of the base. Students observe a variety of prisms, then choose one to construct by cutting and folding the corresponding net.

Prism Dissection: Surface Area

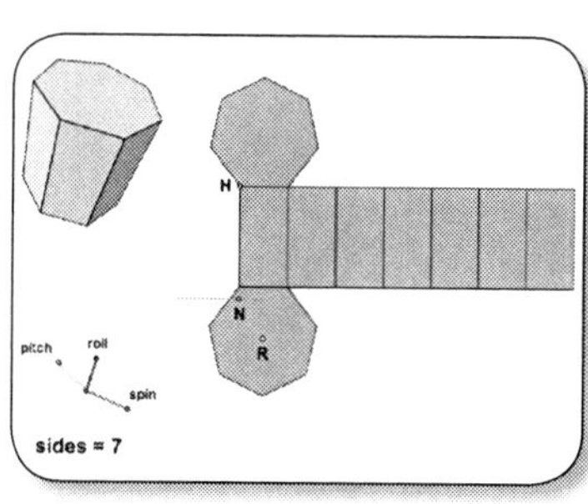

Students find the surface area of a regular right prism using a net that appears next to a three-dimensional view of the prism. They ensure the generality of their results by changing the dimensions and number of sides of the prism as they work. By increasing the number of sides of the base, they relate the surface area of the prism to the surface area of a cylinder, giving them an informal opportunity to think about limits.

Pyramid Dissection: Surface Area

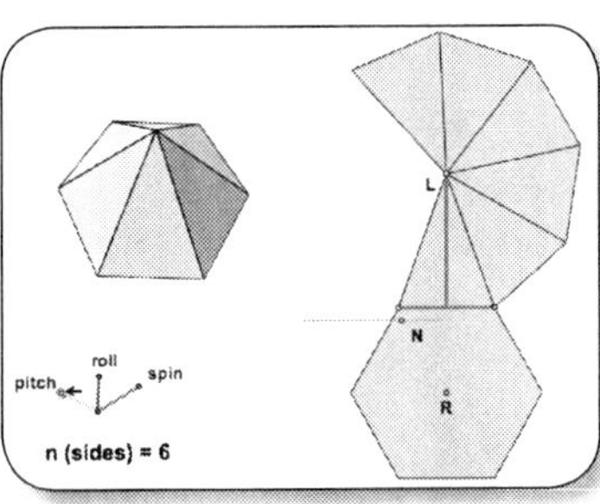

Students find the surface area of a regular pyramid (a pyramid with a regular polygon base) using a net that appears along a three-dimensional view of the pyramid. They ensure the generality of their results by changing the dimensions and number of faces of the base. By increasing the number of faces, students extend their results to the surface area of a cone, giving them an informal opportunity to think about limits.

Stack It Up: Volume of Rectangular Prisms

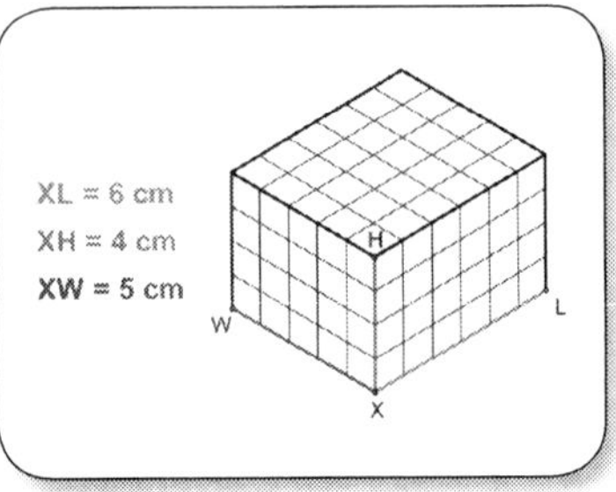

Students develop conceptual understanding of volume of prisms as they manipulate a dynamic model to build boxes filled with layers of cubes. Students use their discoveries to devise a method for finding the volume of a rectangular prism without counting cubes.

Perfect Packages: Surface Area and Volume

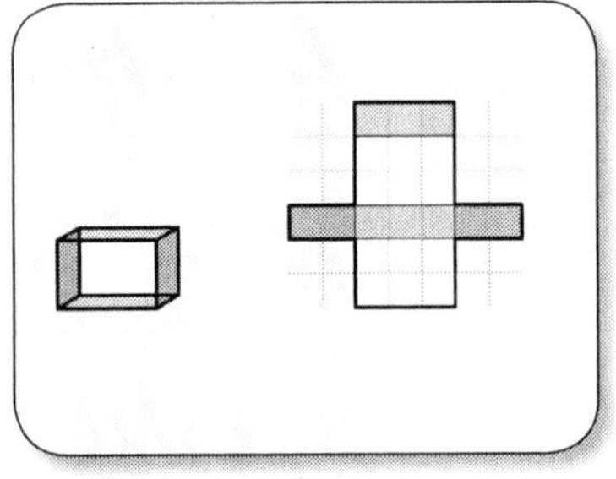

Students explore the relationship between surface area and volume by changing the dimensions of a rectangular prism and its net. Students work to find all surface areas of a rectangular prism having a given volume and explore strategies for finding the least and greatest surface area and volume.

INTRODUCE

Expect to spend 15 minutes for this introduction, which begins with a hands-on experience.

1. Distribute the worksheet. Show students Sample Net 1, and fold it to form a cube.

2. Explain, ***Now it's your turn. Cut out net 1 along the solid lines. Don't cut the dotted lines. After you cut it out, fold on the dotted lines to make a cube. Notice how each number ends up in a particular position on the finished cube. When you finish, answer step 1 on your worksheet.*** Give students time to cut and fold. ***Did everyone's cube come out the same? Let's check. What face is opposite face 1?*** All students should agree that 4 is opposite 1. ***What face is opposite face 2?*** Students should agree that 6 is opposite 2. ***What face is opposite face 3?*** Students should agree that 5 is opposite 3. ***Could you have folded it differently to get different results?*** [No, there's only one way to fold it.] Discuss the answer to this question.

3. ***Now cut out net 2 and fold it into a cube. Write your observations under step 2 on your worksheet.*** Allow enough time for students to experiment and convince themselves that this net does not correspond to a cube. ***Were you able to make a cube from this net? Could anyone make a cube? Why not?*** Different students will have different explanations; be sure to get responses from several students in their own words. Use Sample Net 2 to demonstrate the folding and the impossibility of forming a cube.

4. ***On the last page of the worksheet, draw two nets of your own. Design the first one so that you can fold it into a cube. Design the second so you cannot fold it into a cube. Then cut and fold them to make sure that the first net works and the second doesn't.***

   When you're finished, analyze the nets in step 4 and decide which of them could be folded into a cube. Circle the ones that work.

   Leave plenty of time for students to finish cutting and folding their own nets and to do worksheet step 4.

Allow 10 minutes for demonstrating the page "Practice."

5. ***Now we'll use Sketchpad to match up some cubes and their nets.*** Open **Cube Nets.gsp.** Go to page "Practice." Follow these steps to model using the sketch.

- ***Your first job will be to practice rotating the cube.*** Call attention to the perspective drawing of a cube. Have students observe as you use the *spin*, *pitch*, and *roll* controls. Drag the endpoint of each control to show how the cube rotates around the different axes.
- Students will notice that only two faces of the cube are colored. ***Our challenge is to color the other four faces, using the net as a guide.***
- Rotate the cube so that the blue and yellow faces are visible. You can see the blue and yellow faces. Point to a face that is not colored. ***What color should this uncolored face be? How can you tell from the net?*** Take responses. Demonstrate how to change the color of the uncolored face. Emphasize that students must click the middle of the face, not an edge, to select it.

If the face in question is one of the given faces, then it will not be possible to select it.

- Rotate the cube to bring another uncolored face into view. ***What color should this uncolored face be? How can you tell?***
- If this question is hard, ask students whether it would be easier to rotate in a different direction to bring a different uncolored face into view. Consider showing students how to rotate the cube so that two colored faces of the cube are in the same orientation as the same colors on the net.
- Color the uncolored face. Continuing to work in the same way, color the remaining faces with the class's input.
- Model using the *Wrap* button and the *Toggle* buttons to verify that the colors are correct. First, press the *Toggle* buttons to show what they do. Invite students to explain what the buttons control. Then press *Wrap*. After the net has been wrapped onto the cube, press the *Toggle* buttons again to show that the colors on the net and the colors on the cube match for the three visible faces. Rotate the cube to view other faces, and use the *Toggle* buttons again to check that they are colored correctly. Be sure to confirm the proper matching of all six colors. Tell students that when they are coloring a cube, they should color all the faces before pressing *Wrap*, and that after wrapping they should check to be sure the colors match on all six faces.

DEVELOP

Expect students at computers to spend about 30 minutes on worksheet steps 5–9. You may need to have them split this work between two different sessions, having them pick up at the beginning of the second session where they left off at the end of the first.

6. Tell students where to locate **Cube Nets.gsp.** Have them spend some time on worksheet step 5, experimenting with the controls. Emphasize that each student should get an equal opportunity to experiment.

   Students will get a better feel for the environment by experimenting with no direction. It is unlikely that they will disturb any of the geometric construction. If that does happen, just have them close the document without saving and then open the file again.

7. Once students are familiar with the controls, have them work through worksheet steps 6 and 7.

8. Depending on student progress, now or after worksheet steps 8 or 9 may be an appropriate time to end the first session. Have students save their sketches. Explain that students will save their Sketchpad work so they can continue working on it another day. Demonstrate choosing **File | Save As** and renaming the file, and let students know where to save.

   As you circulate, things to notice are how well students are manipulating the cube with the perspective viewing controls and whether they are having difficulty matching net faces to cube faces. Students may find it easier to figure out the correspondences by taking the time to orient the cube so that the two given colors are in the same relative position (above and below or left and right) on both the net and on the cube.

9. Once students have finished the practice page, have them do worksheet steps 8 and 9.

Corresponding Patterns

This is a good place to start the second session. Project the sketch for use in previewing the next steps with the class. Expect students to spend about 30 minutes on worksheet steps 10–12.

10. Gather the class. Go to page "Pattern 1" of **Cube Nets.gsp.** Explain that now the challenge is to duplicate on each face of the cube the given patterns on the net. To do this, students will draw segments using the five points shown on each cube face. Model drawing segments with attention to these points.

    - A square face can have only one color, but a pattern on a square can be oriented in as many as four different directions. Finding the right pattern is not enough. Students must also have it pointed in the right direction.

- Use **Edit | Line Style** to make your segments thin and **Display | Color** to make them dark. Explain that this way both the new and existing segments will be visible when they are overlaid, and it will be easier to check the work.
- Model by drawing the correct pattern on one face, but orient it incorrectly. Press *Wrap*. Students will observe the problem with the orientation. In this example, the face on the right has the correct pattern, but it's upside down. (The thick original segments are in the shape of an upright V, but the thin drawn segments form an upside-down V.)

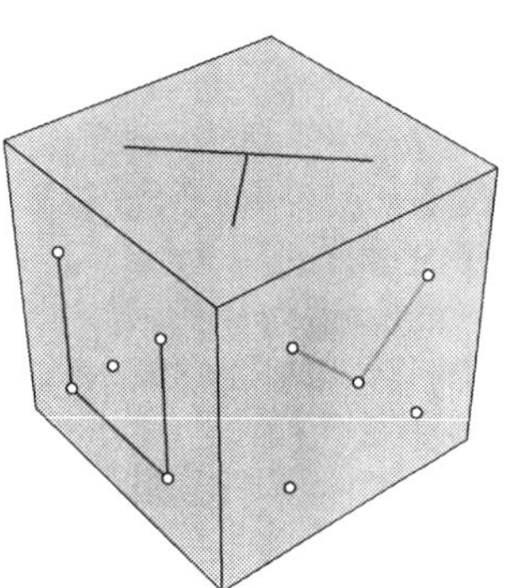

- The colors of the faces are all the same. Explain that students may color-code them if they want to. Color-coding may help them to match the corresponding faces before they make any decisions about the orientations of the patterns.

11. If this is the beginning of the second session, assign students to computers. Students will need their worksheets.
12. As you observe students working, here are some things to note.
    - Be sure students are making their own line segments thin and a dark color.
    - Encourage students to make all of the patterns before pressing *Wrap*. This feature is interesting to watch, but can be a convenient way to avoid thinking.
13. Now or at the end of the Explore More, students should save their sketches so that they have a permanent copy of their work.

Explore More

14. On page "Geodesic," students meet a new challenge. A *geodesic* is a special curve drawn on the surface of a solid. It represents the shortest path between any two points on the surface. When a net of the solid is flattened out, the geodesic becomes a straight line.

    Here, part of the geodesic is drawn on a given net. This particular geodesic crosses each edge only at the midpoint of the edge. Two points are labeled on the cube and on the net.

Students complete the geodesic on the cube. Unlike the other cubes, this one is transparent. Students can use the same strategy they used for the patterned nets.

SUMMARIZE

Allow about 15 minutes for summary discussion.

15. Have students discuss their experiences matching the nets and cubes.

    Did it get easier to color faces? To draw the patterns?

    What strategies did you use to make it easier?

    What do you understand about nets and cubes that you didn't know before?

16. If students had time to explore page "Geodesic," have them discuss their experiences. Focus on these ideas.

    - Did students use the same strategy they used for the patterned nets?
    - What can students say about the shape that is created? Students may conjecture that the shape is a regular hexagon. [It is.] How can they demonstrate this? [They can manipulate the view so that they are looking directly at the hexagon, but students should not be expected to prove congruent sides and angles at this level.]
    - Do students think the shape is a planar (flat) shape? [It is.] If so, can they show that it is? [They can do this by manipulating the view so that the hexagon is viewed on edge.]

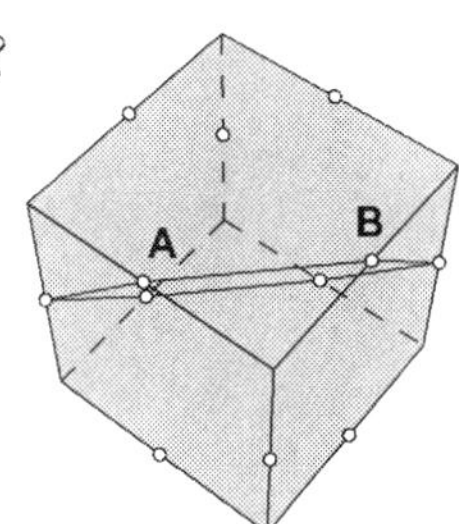

ANSWERS

1. Faces that end up opposite each other are 1 and 4, 2 and 6, and 3 and 5.
2. Net 2 cannot be folded into a cube. Two of the squares always overlap, and one face of the cube is always missing.
3. You can ask students to show you their folded nets as you circulate.

4. 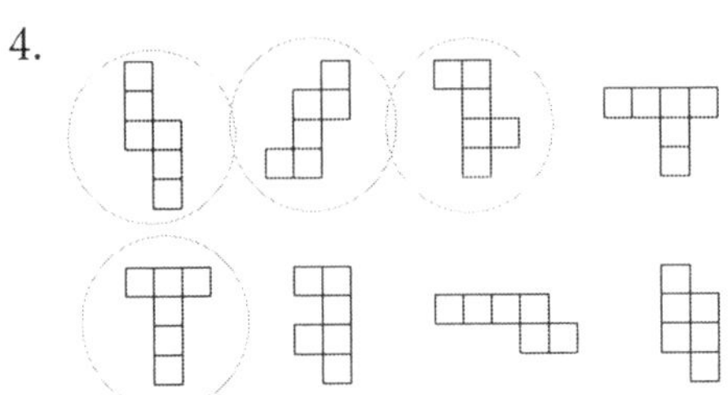

7–12. Students check their own work by pressing *Wrap.*

14. The geodesic is a regular hexagon.

Sample Net 1

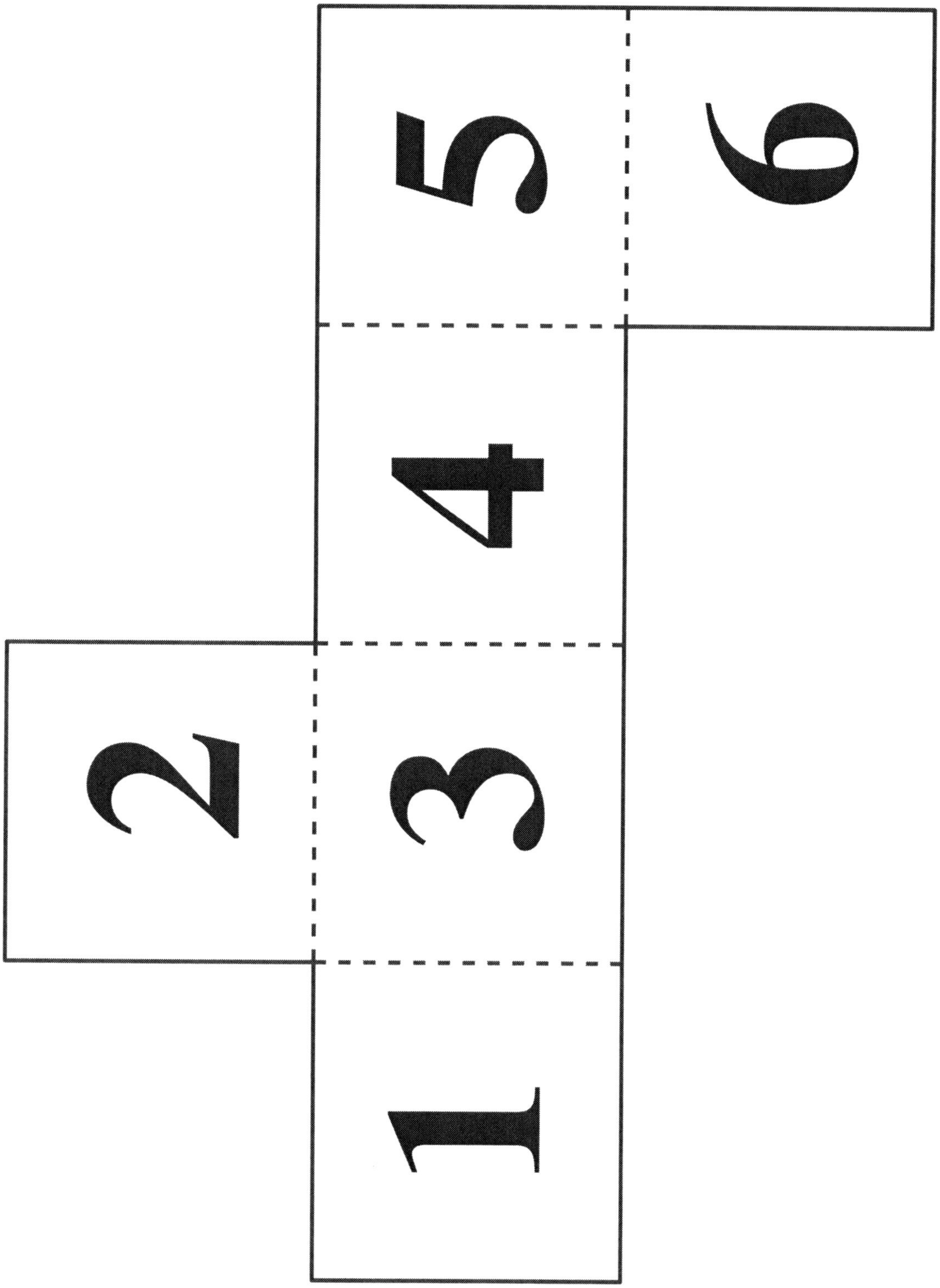

Sample Net 2

Cube Nets

Name:

In this activity you will create matching faces for cubes and their nets.

CUT AND FOLD

1. Cut out Net 1 and fold it into a cube. Notice how each number ends up in a particular position on the finished cube. Which numbers end up on opposite sides from each other?

2. Cut out Net 2 and try to fold it into a cube. What happens?

3. On the last page, draw two more nets, one that can be folded into a cube and one that cannot. Check them by cutting them out and folding them. Sketch the shape of your nets here.

4. Which of the following nets can be folded into a cube, and which cannot? Using only your imagination—no cutting or folding—circle the ones that work.

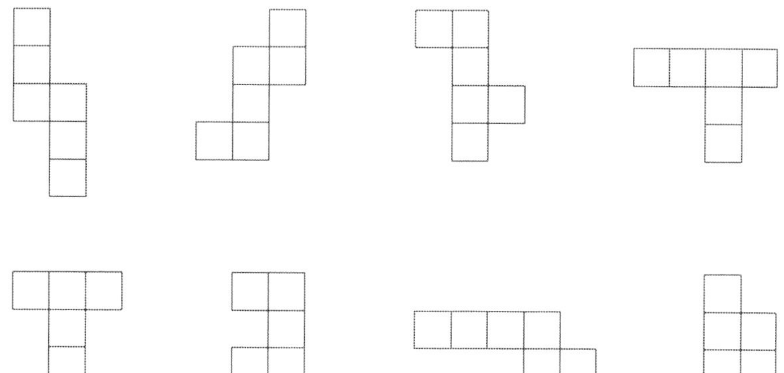

EXPLORE

5. Open **Cube Nets.gsp** and go to page "Practice."

   Use the *spin*, *pitch*, and *roll* controls to turn the cube, so you can look at it from different directions.

6. Color the four blank faces to match the net. To color a face, select it, choose **Display | Color,** and choose the correct matching color.

7. You will check to make sure you colored the cube the same way as the net. First, press *Wrap*. Now, use the *Toggle* buttons to turn the colors on and off for the cube or the net. Be sure to adjust the view so that you check every face.

8. Go to page "Colors 1." Color the cube to match the net. Check your work.

9. Go to page "Colors 2." Here is a different net. This time, the cube faces are colored, but only two of the net faces are. Color the net to match the cube. Then check your work.

10. Go to page "Pattern 1." You will mark the cube faces to match the patterns on the net.

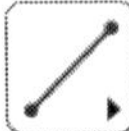

To mark a face, connect the five given points with segments, being careful to draw the pattern in the right direction.

There is only one marked face on the cube. Mark the other five.

11. Press *Wrap* to check your work.

12. Go to pages "Pattern 2," "Pattern 3," and "Pattern 4." Some pages are missing patterns on the cube, and others are missing patterns on the net. Draw the missing patterns on each page. When you have filled in every missing pattern, press *Wrap* to check your work.

13. Describe the strategies you used to match the cube and the net.

EXPLORE MORE

14. Go to page "Geodesic."

A *geodesic* is a special curve drawn on the surface of a solid. It represents the shortest path between any two points on the surface. When a net of the solid is flattened out, the geodesic becomes a straight line.

Here the geodesic is drawn on a net. The labeled points *A* and *B* match one face of the net to one face of the cube.

15. Draw the rest of the geodesic on the cube. What is its shape?

Net 1: Cut out this net and fold it into a cube.

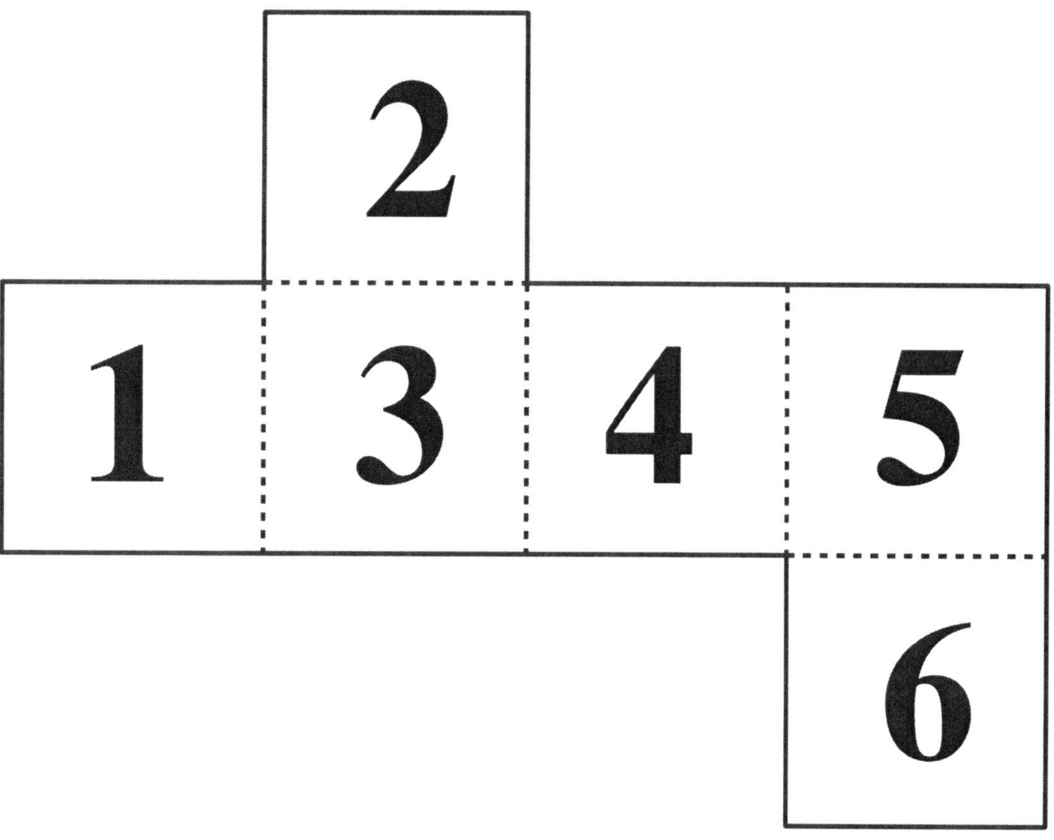

Net 2: What happens when you cut out and fold this net?

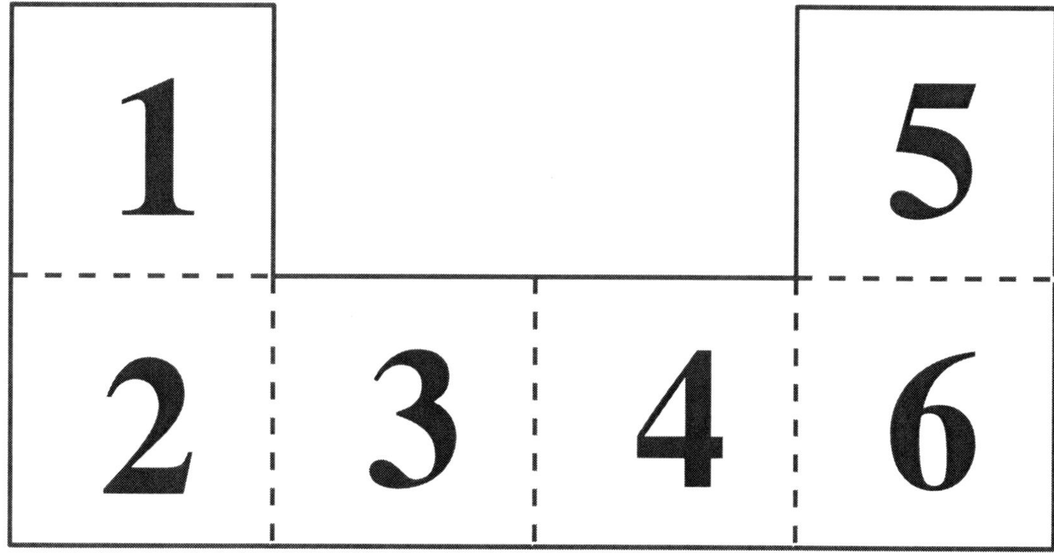

Net 3: Draw a different net you can fold into a cube.

Net 4: Draw a net you cannot fold into a cube.

Prism Nets: Spatial Visualization

ACTIVITY NOTES

INTRODUCE

Expect to spend about 5 minutes.

1. Show students the sample full-page net you cut out, and fold it to form a regular pentagonal prism.
2. Explain, ***This shape is called a* regular right prism. *In this activity you'll experiment with lots of different regular right prisms. You'll look at them from different angles, and you'll change the size of the base, the height of the prism, and the number of sides. You'll use the computer to make it easy to change all these things. After you've looked at a lot of prisms on the computer, you'll print your own net so you can make a prism like this one, but using your own choices for the dimensions and number of sides.***
3. Open **Prism Nets.gsp.** On page "Prism," show students how to manipulate the *spin* control. Tell them, ***I'll leave it to you to experiment with the* pitch *and* roll *controls.*** Then drag the *height* control and tell them, ***I'll leave it to you to experiment with the* radius *control.***
4. Tell students, ***There's also a control for the number of sides.*** Show them how to double-click the parameter *sides* to change its value. Also show them how to select it and use the + and − keys on the keyboard to increase or decrease the value. Point out that they need to hold the Shift key when they press the + sign.

DEVELOP

Expect students to spend about 35 minutes, including 20 minutes at computers and 15 minutes making the prism.

5. Assign students to computers and tell them where to locate **Prism Nets.gsp.** Distribute the worksheet. Tell students to work through step 7 and to be sure to answer the questions in steps 3–7. Give students plenty of time to explore different prisms and to answer the questions. Make sure that all students have a chance to manipulate the viewing controls and the prism controls. Some groups of students will likely go on to worksheet steps 8 and beyond before the rest of the class is ready. Let them do so.
6. Once nearly all students have finished step 6, call the class to attention. Ask them, ***In step 7, you are shown a net. How does the shape of the net relate to the shape of the prism?*** Have students discuss the relationships.
7. Tell them, ***On page "Prism 2," you'll control the prism by changing its net. Go ahead to this page, and use it to answer the questions in steps 9 through 13.***

Give students plenty of time to answer these questions, making sure that all students have a chance to manipulate the nets and the viewing controls.

8. Once most students have finished step 13, call the class to attention. Have them discuss their answers to worksheet steps 12 and 13.
   For step 13, some students may say that the shape of the base becomes more and more circular, and others may observe that the shape disappears when point *N* is dragged all the way to the right. Encourage students to conjecture about what is happening here—why does the diagram disappear?

9. Tell them, ***On page "Prism Net," you'll design your own prism. You'll change the dimensions, the number of sides, and the color. Just make sure the prism stays within the boundaries; if it gets too big, it won't print on a single page. Then you'll print it, cut it, fold it, and tape it into its three-dimensional shape. Remember to record the numbers you used in step 14.***

10. Give students plenty of time to assemble their prisms. Each pair of students should print and assemble a single prism to minimize time spent waiting for the printer. If no printer is available, have students save their sketches, and encourage them to print the nets and assemble the prisms for homework, outside of class.

SUMMARIZE

Allow about 5 minutes.

11. Have students discuss their experiences as they experimented with the nets.

    Did you have trouble connecting the shape of the net with the shape of the prism?

    What helped you understand the connection between the net and the prism?

    These prisms are called regular right prisms. ***Why do you think this name is used? What does the word*** right ***refer to, and what does the word*** regular ***refer to?***

    What happens to the prism as the number of sides increases?

ANSWERS

3. When $n = 3$, the base is an equilateral triangle.

4. When the *pitch* control is horizontal to the left, the top base is all you see. When it's horizontal to the right, the bottom base is all you see. When you're viewing only the base, the *spin* control spins the base around.

5. Student conjectures will vary, and any conjecture is fine. Point R actually determines the distance from the center of the base to each side of the base, measured to the midpoint of the side. As the number of sides increases, this measurement approaches the radius of the limiting circle. Students should begin to see this more clearly as they manipulate point R on page "Prism 2."

6. With 40 sides, the base looks like a circle.

9. Numbers will vary, but for a short wide prism, h should be less than r.

10. For a tall and skinny prism, h should be quite a bit greater than r.

11. The smallest possible number of sides is 3, resulting in a triangular prism.

12. As in step 6, the prism and the bases of the net appear circular when n is large.

13. When N is dragged all the way to the right, the number of sides is shown as infinity, and the prism and net both disappear. Encourage students to discuss why this might happen. Also, encourage them to discuss the largest number of sides they were able to make by dragging N near the right end of the segment without going quite to the end.

14. Answers will vary.

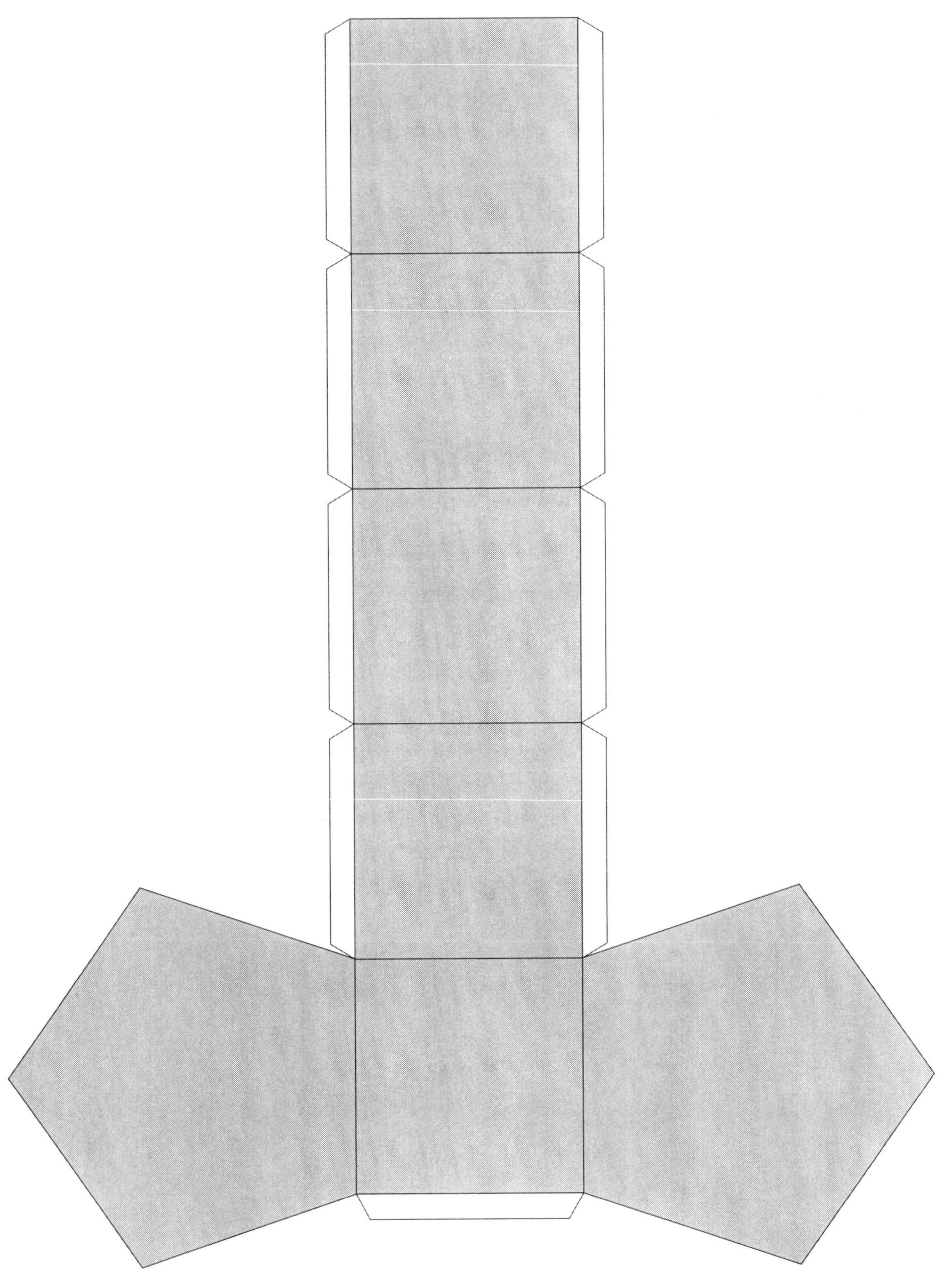

Prism Nets

Name:

In this activity you'll view a three-dimensional shape called a *regular right prism.* You'll change its dimensions and number of sides, and you'll make a prism by folding a two-dimensional net.

EXPLORE

1. Open the sketch **Prism Nets.gsp** and go to page "Prism." Drag the control points labeled *spin, pitch,* and *roll* to view the prism from different angles.

2. Double-click *sides* and change the number of sides. View the new prism from different angles. Then select *sides* and press the + key on the keyboard. Use the + and − keys to view prisms with different numbers of sides. Also drag points *R* and *H* to view prisms with different dimensions.

3. Change the number of sides to 3. What does the base look like now?

4. Adjust *spin, pitch,* and *roll* so you can see only the base. In other words, the base hides the rest of the prism. Which control did you change to make this happen? When you are viewing only the base, what does the *spin* control do?

5. Why do you think the point that determines the size of the base is labeled *R,* even though the base is not a circle? Describe in your own words and show with your own picture what distance this might actually measure.

6. Increase the number of sides to about 40. What does the base look like now?

7. Press *Show Net* to see a two-dimensional net that can be folded into the prism. Change the values for *r, h,* and the number of sides to see how the shape of the two-dimensional net corresponds to the shape of the prism.

8. Go to page "Prism 2." Here you control the prism by dragging points *R, H,* and *N* on the net. Experiment with different values for *r* and *h* and with different numbers of sides. View each prism from different angles.

9. Make the prism short and wide. What numbers did you use for *r* and *h*?

10. Make the prism tall and skinny. What numbers did you use for *r* and *h*?

11. Make the number of sides as small as you can. What happens to the prism?

12. Increase the number of sides to about 30. What does the prism look like?

13. What happens if you make the number of sides as large as you can? Why do you think this happens?

CONSTRUCT

14. Go to page "Prism Net." This page has only a net. Change the values for *r*, *h*, and the number of sides, and measure the height and radius. Also measure the length of one side of the base. (Use the thick red side to make this measurement.) Write down the numbers you used.

15. Adjust the points on the red, green, and blue segments to choose a nice color for your prism net.

16. Make sure the net fits within the gray rectangle, and then press *Hide Controls.* Choose **File | Page Setup** and set the page to print in landscape view. Choose **File | Print Preview.** If necessary, change the scale so the net fits on one page. Then click **Print.**

17. Cut out the net, fold along the lines, and glue or tape to create your three-dimensional shape.

Prisim Dissection: Surface Area

INTRODUCE

Project the sketch for viewing by the class. Expect to spend about 5 minutes.

1. Show students the sample full-page net you cut out, and fold it to form a regular pentagonal prism. Explain, ***Today you will find the surface area of regular right prisms. You'll start with a pentagonal prism and then figure out a formula that works for other regular right prisms.***

2. Open **Prism Dissection.gsp.** Explain, ***On page "Prism," you'll review how to change the viewing angle and how to change the dimensions and number of sides.*** Model the use of the *spin*, *pitch*, and *roll* controls and how to change the number of sides.

3. ***What does surface area mean?*** Review the definition with the class. Here is a sample definition: The surface area of a prism is the combined area of all the faces of the prism. ***If you want to find the surface area of a prism, how can a net help you?*** Encourage students to explain why the area of the net is the same as the surface area of the prism itself.

4. If students have not previously used the Construct or Measure menus, briefly demonstrate how to construct a midpoint and how to measure the distance between two points. If students have not previously used the Calculator, show them how to choose **Number | Calculate** and how to click an object in the sketch to enter it into a calculation.

DEVELOP

Expect students at computers to spend about 30 minutes.

5. Assign students to computers and tell them where to locate **Prism Dissection.gsp.** Distribute the worksheet. Tell them, ***Once you've reviewed the controls on page "Prism," your next job will be to go to page "Base 1" and find the area of one base. If you can, try to figure out the measurements and calculations you need without using the hint.***

6. Give students time to work on their measurements and calculations in worksheet steps 5 and 6. If necessary, remind them to enter existing measurements into the Calculator by clicking on them rather than by typing the numbers in. Some students may be ready to go on to page "Base 2" before others have finished their calculations. If they do so, you can check their results for step 6 to make sure they answered 3.63. (All students should get this same result for a pentagonal base, no matter what size it is.)

7. When most students have finished worksheet step 6, call the class to attention and ask several students to report their values of *r*, the length of a side, and the step 6 calculation. Ask, ***What's interesting about these results from different-size bases?*** There's no need at this point for students to explain why the ratios are all the same, but it is important to bring this fact to their attention.

8. Tell students, ***If you used the number 5 in your calculation of the area, your calculation will work only for bases that are pentagons. On page "Base 2," do the same calculation, but in a way that will work for any polygon, no matter how many sides it has.***

9. Give students time to work on worksheet steps 7–11. Some students may be ready to go on to page "Faces" before others have finished their calculations. If they do so, you can check their results for step 10 to make sure they answered 3.14.

10. When most students have finished worksheet step 11, call the class to attention and ask several students to report their values of *r*, *n*, and the step 10 calculation. Ask, ***What's interesting about the calculated ratio now?*** Some students may mention that not only are the ratios very close to equal, but they get closer and closer to π as the number of sides gets larger and larger. Don't discuss this result in detail yet; give students time to think about it while they do the next few steps of the worksheet.

11. Tell students, ***Now that you've calculated the area of a base, you also need to calculate the area of a lateral face of the prism. Go on to page "Faces" and then to page "Area" when you're ready. If you finish early, try some of the Explore More questions.***

12. Give students time to work on worksheet steps 12–16. You might also consider assigning some or all of the Explore More, depending on your curriculum needs. If so, allot additional time.

SUMMARIZE

Project the sketch. Expect to spend about 10 minutes.

13. Have students discuss their results. Here are some questions for discussion.

    How did you find the area of the base?

    How did the net help you to understand the problem of finding the surface area?

You ended up with a formula that involved the values of n, r, h, and the length of a side of the polygon. Do you really need all of these values?

Encourage students to realize that these are interdependent. At the middle school level, you might discuss that for a given number of sides, the side length increases in proportion to the value of r. As you drag R, observe the side length and note that the triangles used to find the area of the base remain similar. At the high school level, you can use trigonometry to explicitly relate the side length to the values of r and n.

What interesting things happen as n increases? Students will observe that the shape becomes more and more like a cylinder. ***What happens to the area of the base?*** Some students will have noticed that the ratio in worksheet step 10 is 3.14, and that as the number of sides increases, the ratio approaches π. Some also may have drawn the conclusion that the area of the polygon approaches πr^2, the area of a circle. ***What happens to the area of the lateral faces?*** Some students will have noticed that the ratio in worksheet step 15 approaches 2π. Some may also have drawn the conclusion that the area of the lateral faces approaches $2\pi rh$, the circumference of the circle multiplied by the height. (Explore More worksheet step 18 explicitly asks students to use these facts to develop a formula for the surface area of a cylinder.)

ANSWERS

3. Answers will vary.

For the area calculations in steps 5 and 8, students can partition the base into 5 triangles with bases that are the side length (represented by s here) or 10 triangles with bases that are half the side length (represented by t here).

5. $A = 5 \cdot (s \cdot r)/2$ (alternatively, $A = 10 \cdot (t \cdot r)/2$)
6. For a five-sided polygon, the ratio is 3.63 and does not depend on the size.
8. $A = n \cdot (s \cdot r)/2$ (alternatively, $A = 2n \cdot (t \cdot r)/2$)
9. The value of the ratio depends on n. Any two students using the same value of n should have the same ratio.

10. For approximately $n = 100$, the ratio is 3.14. As n becomes large, the ratio appears to approach π. This makes sense because the more sides the polygon has, the more it looks like a circle, and the closer the measurement r is to the radius of the circle. Because the area of a circle is πr^2, the ratio of the area to r^2 is π.

11. It makes sense to use r because r measures the radius of the limiting circle. In fact, when n is small, r is the radius of the inscribed circle, though students may not make this observation specifically.

13. The area of the lateral faces is $n \cdot s \cdot h$.

14. When you divide by $r \cdot h$, the resulting value changes with n, but not with r or h.

15. When the value of n is large, the ratio becomes approximately 2π. One explanation is that the face area is equal to the perimeter $(n \cdot s)$ multiplied by h. As n increases, the perimeter approaches the circumference of a circle, so the face area approaches $2\pi r$ multiplied by h. When this value is divided by rh, the result is 2π.

16. The total surface area of the prism is the sum of the base areas and lateral face areas, so it can be expressed as $A = 2(n(s \cdot r)/2) + n(s \cdot h) = n \cdot s \cdot r + n \cdot s \cdot h$. The product $n \cdot s$ is the perimeter of the base (p), so the formula could be written as $n = p \cdot r + p \cdot h$.

18. When n becomes large, $n \cdot s$ approaches $2\pi r$, so the area approaches $2\pi r^2 + 2\pi rh$.

19. The volume must equal the area of the base multiplied by the height: $V = n \cdot s \cdot r \cdot h$, approaching a limit of $V = \pi r^2 h$.

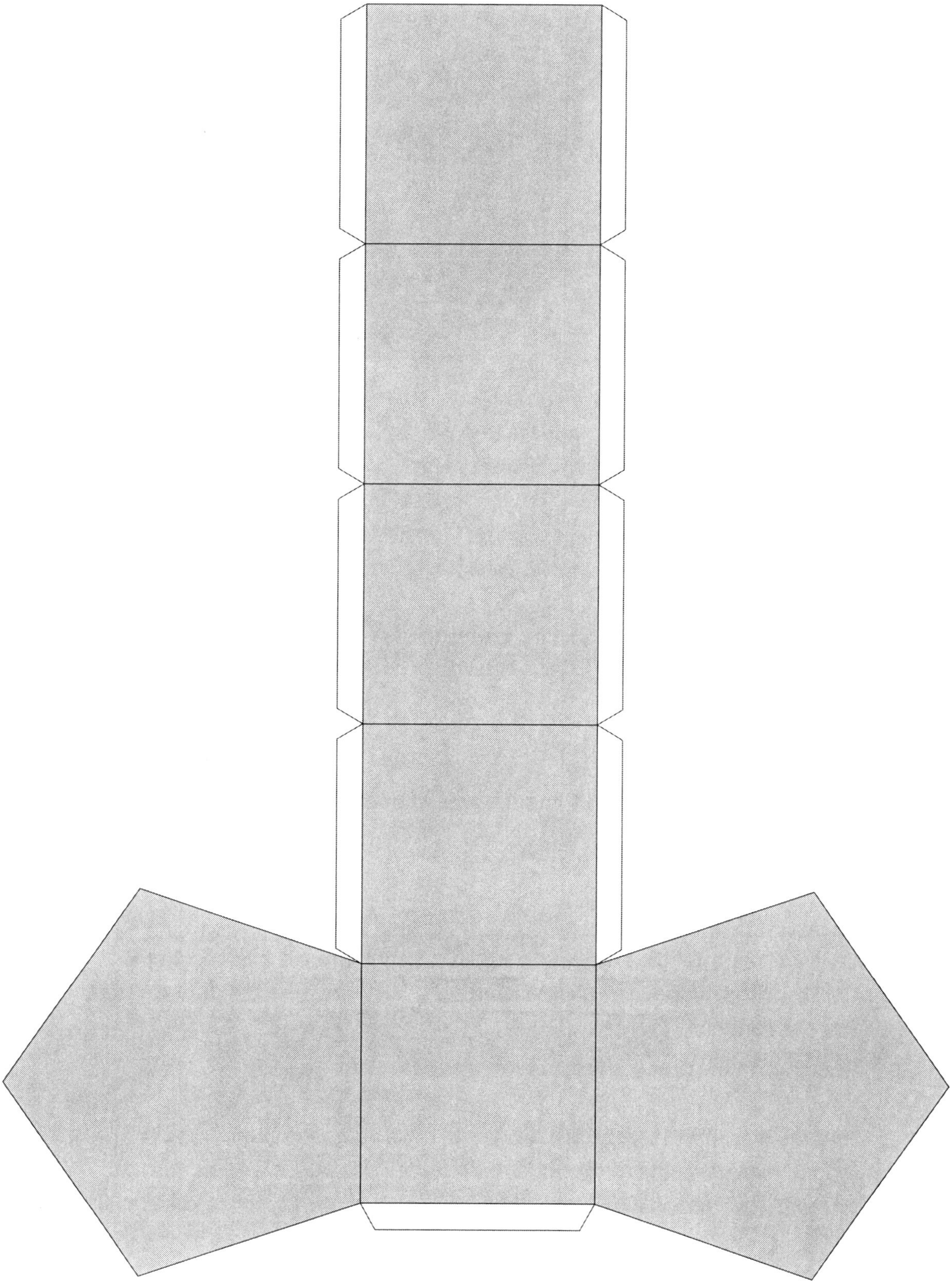

Prism Dissection

Name:

In this activity you will create a regular prism with your choice of the height, the number of sides, and the size of the base. Then you'll calculate its surface area.

EXPLORE

1. Open **Prism Dissection.gsp** and go to page "Prism." Drag *spin, pitch,* and *roll* to view the regular prism from different angles.

2. Drag *N* to increase the number of sides. View the new prism from different angles. Drag *R* and *H* to change the shape of the prism.

3. How does the two-dimensional net correspond to the shape of the prism?

To find the surface area, you must find the area of both bases and of the lateral faces.

Base Area

4. On page "Base 1," the base has five sides. Change the size by dragging *R*. Construct the midpoint of the thick red side. Then measure the distance from *R* to the midpoint.

Double-click the distance measurement and label it *r*.

5. Find the area of the base. You'll divide the regular pentagon into simpler shapes, measure some distances, and do some calculations. (If you're stuck, pressing *Show Hint* can give you some ideas.) Write down your calculation and result.

Prism Dissection

continued

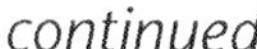

6. Divide your resulting area by r^2. What value do you get? How does it compare with the results of other students?

7. On page "Base 2," measure distance r as you did in step 4. Change the number of sides and the size of the base.

8. Using n to represent the number of sides, write a calculation for the area of the base that will be correct for any value of n. Write down the result for your value of n.

9. Divide the area by r^2. What value do you get?

10. Increase the number of sides to more than 50. Now what is the value of the area divided by r^2? Have you seen this number before? Why do you think you get this value?

11. Why does it make sense to use the letter r to stand for the distance from the center to a side of the base?

Face Area

12. On page "Faces," drag H, R, and N to change the shape of the prism and the number of sides. Measure the distance r as you did in step 4. Then measure the height of the prism, and label it h.

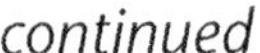

13. Find the total area of the lateral faces. Write a calculation that will be correct for any value of *n*. Write down the result for your value of *n*.

14. Divide your resulting area by the product of *r* and *h*. What value do you get? Does this value change if you drag *R* or *H?* Does it change if you drag *N?*

15. What is this value when the number of sides is at least 50? Why?

Total Area

16. On page "Area," do the necessary constructions, measurements, and calculations to find the total area of the regular prism, including both bases and the lateral faces. Write down your calculation and result.

EXPLORE MORE

17. On page "Net," set *H, R,* and *N* to match the prism you made on page "Area." Choose **File | Page Setup,** set the page to print in landscape view, and then choose **File | Print Preview.** If necessary, change the scale so the net fits on one page. Then click **Print.** Cut out the net, fold along the lines, and glue or tape your three-dimensional prism together. Label each base and face with the area you calculated based on the measurements.

18. What does the prism look like when the number of sides is very large? How could you calculate the approximate surface area of this shape without using the value of *n?* (Your answers to steps 10 and 15 may be useful.)

19. Use your measurements to calculate the volume of the prism. Explain why you used the calculation you did. Then write your method as a formula.

20. On page "Explore More," you can experiment with the advanced controls that affect the look of this three-dimensional model.

Pyramid Dissection: Surface Area

INTRODUCE

Project the sketch for viewing by the class. Expect to spend about 5 minutes.

1. Students should have already completed the activities Prism Nets and Prism Dissection. Show students the sample full-page net you cut out, and fold it to form a regular pentagonal pyramid. Explain, ***Today you'll use Sketchpad to find the surface area of regular pyramids. You'll start with a pentagonal pyramid and then figure out a formula that works for other regular pyramids.***
2. Open **Pyramid Dissection.gsp.** Explain, ***On page "Pyramid," you'll review how to change the viewing angle and how to change the dimensions and number of sides.*** If students are not already familiar with the controls, model the use of *spin, pitch,* and *roll,* and the use of *N, R,* and *L.*
3. ***To find the surface area of a pyramid, what would you have to measure?*** Some students might provide a general description that you must measure the areas of the base and the five triangular lateral faces. Others might give more detail, describing how to measure the base and height of each triangle. ***How can a net help you?*** Students should see that the area of the net is equal to the surface area of the pyramid.
4. Holding up the folded pyramid, ask, ***Where would you measure the height of this pyramid?*** Encourage students to notice that there are two possible height measurements: the vertical distance from the center of the base to the vertex (height of the pyramid) and the distance from the base of a lateral face to the vertex (height of a triangle). Encourage students to discuss how these two different measurements might be useful, and discuss with them why it's important to avoid confusion by distinguishing the two heights. You may want to ask them to propose their own names for these measurements. Explain, ***On your worksheet, the distance from the base of a triangular face to the vertex is called the*** **slant height** ***and is labeled l.***
5. Students should have previous experience with the Construct and Measure menus. You may want to briefly review how to construct a midpoint, how to measure the distance between two points, how to change the label of a measurement, and how to click on an object in the sketch to enter it into the Calculator.

DEVELOP

Expect students at computers to spend about 30 minutes.

6. Assign students to computers and tell them where to locate **Pyramid Dissection.gsp.** Distribute the worksheet. Tell them, ***First review the controls on page "Pyramid" and answer the questions in steps 3 and 4. Then you'll go to page "Base" and find the area of the base. Try to figure out the measurements and calculations you need without using the hint.***

7. Give students time to work on their measurements and calculations. If necessary, remind them to enter existing measurements into the Calculator by clicking on them rather than by typing in the numbers. Some students may be ready to go on to page "Faces" before others have finished their calculations. If they do so, you can check their results for worksheet steps 6, 7, and 9. For step 9, make sure they answered approximately 6.29 for the ratio of the perimeter to r, and approximately 3.14 for the ratio of the area to r^2. Also make sure they've written an explanation for why they got these values.

   If students have different answers for worksheet step 9, check to make sure that they used the value of n and the actual measurements rather than typing in numbers, so that their results are correct no matter how they change the number of sides.

8. When most students have finished worksheet step 9, call the class together and ask several students to report their measurements and calculations of perimeters and areas from worksheet steps 6 and 7. Ask, ***What values did you get for the two calculations in step 9?*** Encourage students to explain why the two ratios come out to 2π and π. You may want to ask them whether the values are exactly 2π and π, and if not, why not? How could they measure the percentage by which the ratios differ from 2π and π?

9. Ask students, ***When you did your calculations, why was it important to click on the measurements in the sketch rather than just typing in the number?*** Encourage students to observe that they want their calculations to be correct even when they change the pyramid dimensions or the number of sides.

10. Tell students, ***Now you'll calculate the area of the lateral faces of the pyramid. Go on to page "Faces" and then to page "Area" when you're ready. If you finish early, try some of the Explore More problems.***

11. Give students time to work on worksheet steps 10–13. In step 12, some students may need a hint; you can suggest that they see what happens if they divide their numeric answer by rl.

    Consider assigning some or all of the Explore More questions, depending on your curriculum needs. If you do so, allot additional time.

SUMMARIZE

Project the sketch. Expect to spend about 10 minutes.

12. Have students discuss their results. Here are some questions for discussion.

    How did you find the area of the base?

    How did the net help you to understand the problem of finding the surface area?

    You ended up with a formula that involved the values of n, r, l, and the length of a side of the base. Do you really need all of these values?

    Encourage students to realize that these are interdependent. At the middle school level, you might discuss that for a given number of sides, the side length increases in proportion to the value of r. As you drag R, observe the side length and note that the triangles used to find the area of the base remain similar. At the high school level, you can use trigonometry to relate the side length explicitly to the values of r and n.

13. ***What interesting things happen as n increases?*** Students will observe that the shape becomes more and more like a cone. ***What happens to the area of the base?*** Students will have noticed that the ratio of area to r^2 in worksheet step 9 approaches π. They should be ready to draw the conclusion that the area of the polygon approaches πr^2, the area of a circle. ***What happens to the total area of the lateral faces?*** Some students will have noticed that the ratio of the perimeter to r in worksheet step 9 approaches 2π. Encourage students to explain how they could use the perimeter to make it easy to calculate the sum of the areas of these faces, by calculating *perimeter* × *slant height*/2. (Explore More worksheet step 15 explicitly asks students to use these facts to develop a formula for the surface area of a cone.)

EXTEND

What other questions might you ask about pyramids? Encourage all inquiry. Here are some ideas students might suggest.

What do pitch, roll, and spin really do? Are there other similar movements?

Why are the circumference and the area of a circle given by the familiar formulas? Is this activity a proof?

Can you do something like this to find the volume of the pyramid or cone?

Can you get the surface area of other curved figures, like a sphere, by taking some figure and increasing the number of faces?

ANSWERS

3. Student answers will vary. The *pitch* control allows you to see a top view, so that the pyramid looks like a regular polygon. In this view the *spin* control rotates the polygon about its center.

4. You cannot make the slant height l smaller than the value of r because the slant height must reach at least from the edge of the base to the center. When the slant height is equal to r, the pyramid is completely flat, with a height of 0. When the slant height is large compared to the radius, the pyramid is tall and skinny.

6. If students use s for the length of one side of the base, the perimeter is ns. Numeric results will vary and should change as students manipulate the dimensions.

7. One way to measure the area of the base is to think of it consisting of n triangles (as suggested by the hint), to measure the base (s) and height (r) of one triangle, and then use $A = sr/2$ to find its area. To find the area of the entire base, students can multiply by n, with the result that the total area is given by $nsr/2$.

9. The ratio of the perimeter to r is approximately 6.29—nearly 2π—and the ratio of the area to r^2 is approximately 3.14—nearly π. When $n = 60$, these values are accurate to about two decimal places.

11. The area of one lateral face is $sl/2$, and the sum of the areas of all these faces is $nsl/2$. If students use the perimeter p in their calculations, they will use the formula $pl/2$. Numeric answers will vary.

12. When the number of sides is large ($n \geq 50$), the perimeter approaches $2\pi r$, so the area of the sides approaches πrl.

13. The total area of the pyramid is the sum of the area of the base and the lateral faces: $nsr/2 + nsl/2$. Students may factor this to write it as $ns(r + l)/2$. Numeric results will vary.

15. When the value of n is large, the base becomes very nearly a circle, and the base area can be written as πr^2. The sum of the areas of the lateral faces approaches πrl, so the surface area of a cone is given by $\pi r^2 + \pi rl$, or $\pi r(r + l)$.

16. The height h of the pyramid (measured from the center of the base to the vertex) and the distance r form two legs of a right triangle, with the slant height l forming the hypotenuse. By the Pythagorean Theorem, $r^2 + h^2 = l^2$. If $l = 15$ cm and $r = 9$ cm, $h = 12$ cm. If $h = 5$ cm and $r = 12$ cm, $l = 13$ cm.

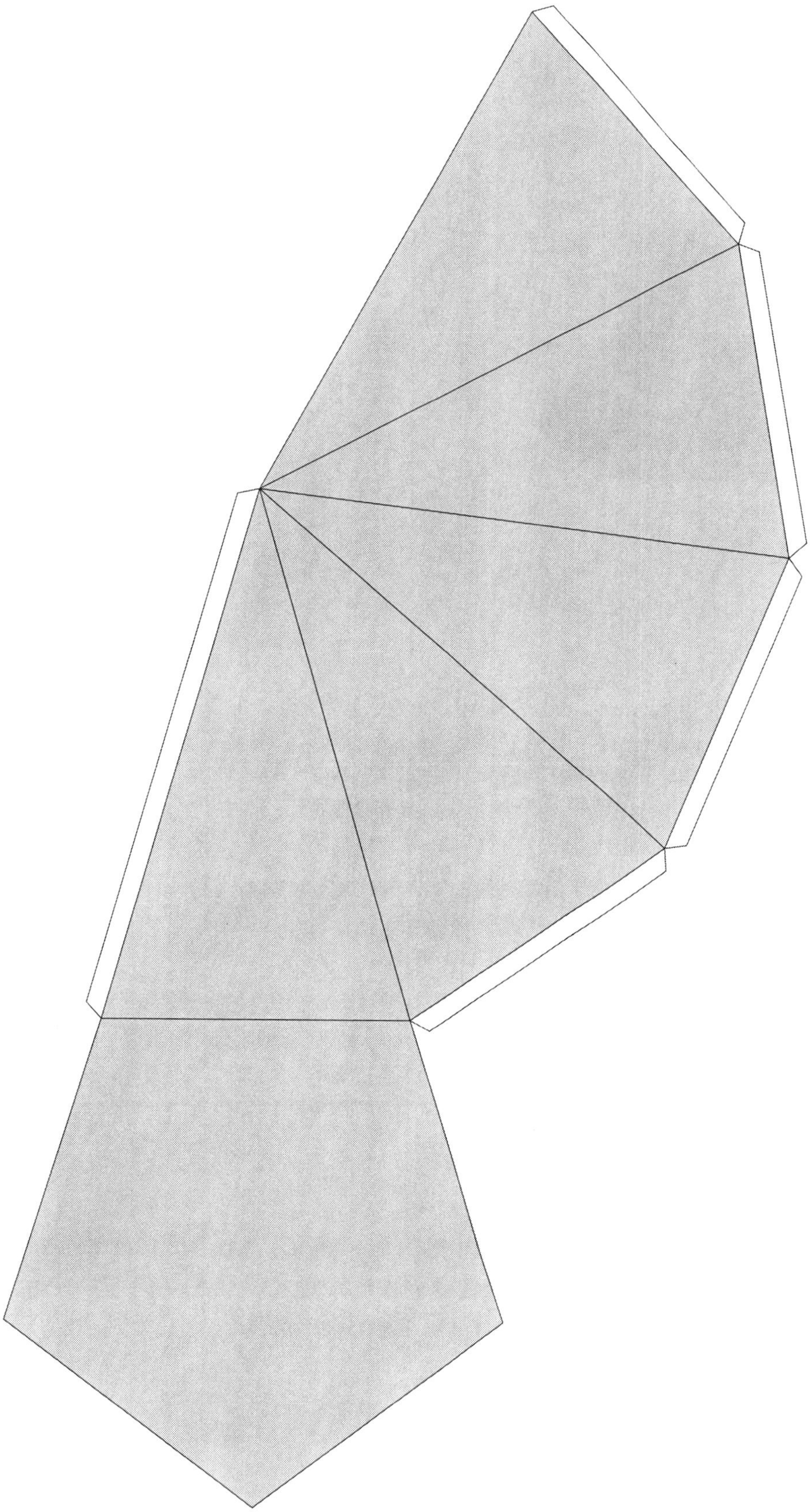

Pyramid Dissection

Name:

In this activity you'll create a regular pyramid with your choice of the height, the number of sides, and the size of the base. Then you'll calculate its surface area.

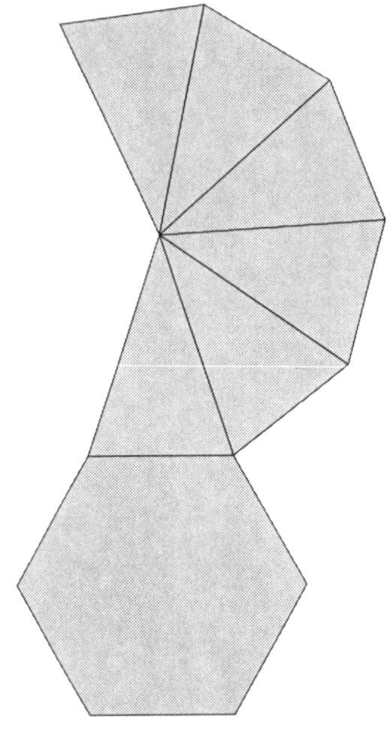

EXPLORE

1. Open **Pyramid Dissection.gsp** and go to page "Pyramid." Use *spin, pitch,* and *roll* to change your view of the regular pyramid.

2. Change the number of sides (N), the size of the base (R), and the slant height (L). View the pyramid from different angles.

3. Adjust the controls to look at the pyramid from directly above. Which control did you use to do this? What does the pyramid look like in this position? What does the *spin* control do now?

4. Explore the shape of the net for various pyramids. How small can you make the slant height? What do the pyramid and the net look like then? What happens if you make the slant height large and the radius small?

To find the surface area, you must find the area of the base and of the lateral faces.

Base Area

5. Go to page "Base." Change the size of the base and change the number of sides. Measure the distance from *R* to the midpoint of the thick red side. Label the measurement *r*.

6. Measure one side of the base and have Sketchpad calculate the perimeter. Write down your calculation and result.

7. Find the area of the base. Imagine dividing the regular polygon into simpler shapes and do some measurements and calculations. (If you're stuck, press *Show Hint* to get some ideas.) Write down your calculation and result.

8. Change the number of sides and the size of the base, and make sure that your perimeter and area calculations seem reasonable. If not, fix them so that they work correctly for any base.

9. Increase the number of sides to more than 50. Divide the perimeter by *r* and divide the area by r^2. Have you seen these two numbers before? Why do you think you get these values?

Face Area

10. On page "Faces," drag *L, R,* and *N* to change the slant height, the size of the base, and the number of sides. Measure the distance *r* and side length *s* as you did in steps 5 and 6. Measure the slant height of the pyramid and label it *l*.

11. Find the total area of the lateral faces. Calculate a value that will be correct for any value of n. Write down the calculation and the result.

12. What is this value when the number of faces is at least 50? Why?

Total Area

13. On page "Area," do the necessary constructions, measurements, and calculations to find the total surface area of the regular pyramid, including the base and the lateral faces. Write down your calculation and result.

EXPLORE MORE

14. On page "Net," set L, R, and N to match the pyramid you made on page "Area." Choose **File | Print Preview** and make sure the net fits on one page. If necessary, click Scale To Fit Page, and then click Print. Label the base and each face with the area you calculated based on the measurements. Cut out the net, fold along the lines, and glue or tape your three-dimensional pyramid together.

15. What does the pyramid look like when the number of lateral faces is large? How could you calculate the surface area of this shape without using the value of n? (Your answers to steps 9 and 12 may be useful.)

16. The *height* (h) of a pyramid is defined as the vertical distance from the base to the *vertex*. In step 10, you measured the slant height l (the height of one of the lateral faces). How can you find the vertical height of the pyramid if you know l and r? (For instance, if $l = 15$ cm and $r = 9$ cm, what is h?) How could you find l if you know h and r? (For instance, if $h = 5$ cm and $r = 12$ cm, what is l?)

17. On page "Explore More," you can experiment with the advanced controls that affect the look of this three-dimensional model.

INTRODUCE

Project the sketch for viewing by the class. Expect to spend about 15 minutes.

1. If centimeter cubes are available, distribute about 25 (at least 20) to each student. Open **Stack It Up.gsp.** Go to page "Layers." ***You see one cube.*** Drag point *L* to add one cube at a time until there are five cubes.

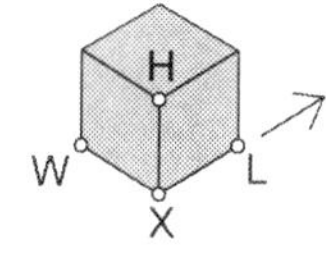

   What is happening? Introduce the language, *row* of cubes, if students don't. For students who need help visualizing the two-dimensional representation as a three-dimensional shape, building this row with centimeter cubes will be helpful.

2. Drag point *W* so that the row of cubes grows to a layer that has 2 rows of 5 cubes, then 3 rows of 5 cubes, and finally 4 rows of 5 cubes. If students have centimeter cubes, have them represent the growing layer using the cubes. With the addition of each row, ask students what is happening. By focusing on the growth pattern, you'll help students to interpret what they see and to develop a deep understanding of volume.

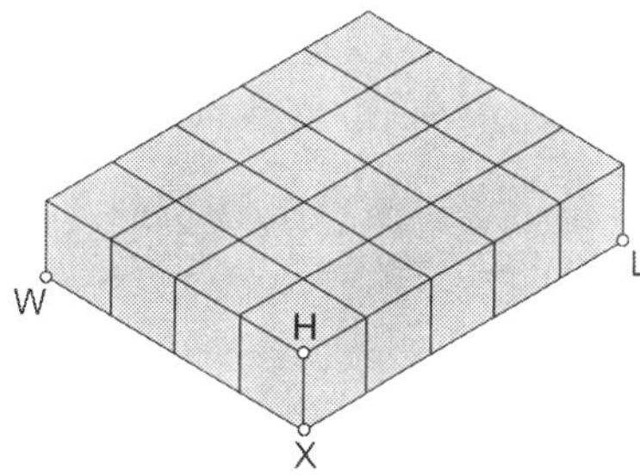

3. ***Let's call this one layer. How many cubes fill this layer? How do you know?*** Elicit the idea that there are 4 rows of 5 cubes, or $4 \times 5 = 20$ cubes. Because there is only one layer, some students may confuse the answer, 20 cubes, with the measure of the *area* of the top face of the layer. Clarify that 20 cubes tells about the amount of space filled, while area is a measure of the amount of space covered. The area of the top face of the layer is 20 square centimeters. Cut a 4-by-5 rectangle from centimeter graph paper and ask whether this is equal to the space filled by the cubes in the model. [It isn't.]

4. Drag point H so that the single layer of cubes grows to 2 layers of 20 cubes each, and then to 3 layers of 20 cubes each. Ask students what is happening.

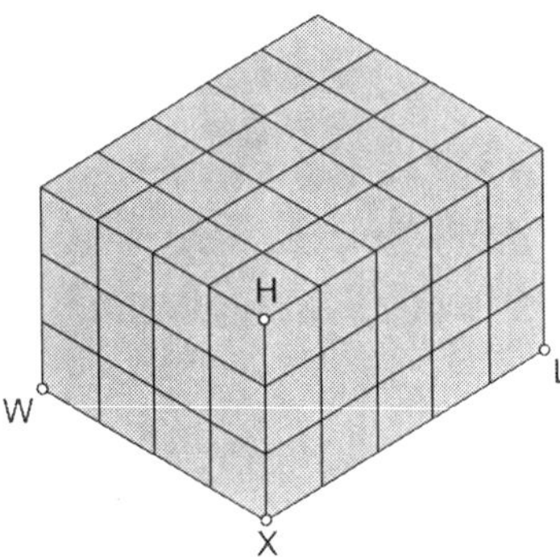

More layers are showing up. They are stacked on top of each other.

The first layer is being copied.

Dragging point H increases the height of the box.

Do not introduce the formula for volume of rectangular prisms (Volume = length × width × height).

How many cubes fill this box? How do you know? Students may explain that there are 3 layers of 20 cubes, or $3 \times 20 = 60$ cubes.

5. Distribute the worksheet. Tell students to work through steps 1–6 and do the Explore More if they have time. ***In each step, you will use the model to make boxes with a certain number of cubes.*** Model enlarging the window so it fills most of the screen. Drag point X and let students know they can move a box to make room to build large boxes. ***The largest box that can be built using the model is 12 cm long, 12 cm wide, and 12 cm tall.***

6. If students will save their work, model choosing **File | Save As** and tell how they should name their files and where to save them.

DEVELOP

Expect students at computers to spend about 30 minutes.

7. Assign students to computers and tell them where to locate **Stack It Up.gsp.** Encourage students to ask a neighbor for help using Sketchpad if needed.

8. Let pairs work at their own pace. As you circulate, observe and listen to students' conversations. Here are some things to notice.

   - In worksheet step 1, students will probably take note of the values for *XL, XW,* and *XH* that appear and update automatically in the sketch. Let students explore these on their own. Some students may begin to make sense of the displayed values and use them; other students may disregard them.

   - In worksheet step 1, some students may not realize that a box with a single row of cubes is one possible solution; they may think that all three dimensions need to be larger than 1 cm.

   - In worksheet steps 1 and 2, challenge students to make a box with no side of length 1 cm. Also notice any students who make the conjecture that the lengths of *XL, XW,* and *XH* must all be factors of the total number of cubes.

   The largest box allowed by the model is 12 cm in each dimension. Thus, students cannot build a 1 cm × 1 cm × 24 cm box.

   - In worksheet steps 5 and 6, students are asked to build two different boxes using the same number of cubes in each box. It's fine if students make boxes that share one dimension, such as a 2 cm × 4 cm × 3 cm box and 2 cm × 1 cm × 12 cm box. Prompt those students to extend their thinking by saying, ***I wonder if it's possible to make two boxes so that they have no lengths in common.***

   - In the Explore More steps, students are given several new problem types. Encourage students to persevere on their own to try to find solutions. Let students know they will hear others' solutions during the class discussion that follows.

Project the sketch. Expect to spend about 30 minutes.

9. If students will save their work, have them do so now.

SUMMARIZE

10. Gather the class. Students should have their worksheets. ***We've been calling our Sketchpad model a box. Mathematicians call it a* rectangular prism.** Review the properties of rectangular prisms.

11. Tell the class that the amount of space inside a three-dimensional shape is called the *volume* of the shape. Another way to say this is that volume tells the amount of space a three-dimensional shape takes up. Students have found the volume of boxes by finding the number of cubes that fill the boxes.

Have students practice using the terms *volume* and *cubic centimeter* throughout this discussion.

Go to page "Build a Box" and show one cube. (Students can examine centimeter cubes as well if they have them.) ***The unit of measure you have been using is called a* cubic centimeter. *Each edge of this cube is 1 centimeter in length. Each face has an area of one square centimeter.*** Drag points to build several small boxes, and ask students to tell the volume of each box expressed in cubic centimeters.

12. Facilitate discussion of methods students have found for determining the volume of rectangular prisms. Begin by saying, I'm thinking of a box that has 2 rows of 10 cubes in the bottom layer, and 5 layers. Write this information on the board.

    How can I figure out the volume of the box? The discussion should bring out the ideas expressed in the student responses that follow.

    First, you have to figure out the number of cubes in a layer. Then multiply that by the number of layers. The box has 2 × 10 cubes in a layer; that's 20. Times 5 is 100. So the volume of the box is 100 cubic centimeters. This is the same as the way we made boxes with the model. First, you make the bottom layer; then you make more layers that have the same number of cubes.

    To find the number of cubes in a layer, just find the area of the bottom of the box. I think about each cube in the bottom layer sitting on one square centimeter, so the number of square centimeters in the area of the bottom of the box is equal to the number of cubic centimeters in the layer. The area of the bottom of your box is 2 × 10. Multiply that by the height, 5, which tells you the number of layers.

    Dragging point L creates the length of a box and dragging point W creates the width of a box. We can find the area of the bottom layer by multiplying the length (L) times the width (W). H is the height of the box. After we find the area of the bottom, we multiply by the height.

13. Discuss the Explore More problems, worksheet steps 7–9. In step 9, students may use this reasoning: The largest amount of cubes Roberto can use is 90 (the largest multiple of 5 that is less than 93); 20 layers would require 100 cubes, so 90 cubes will fill two layers of 5 fewer, or 18 layers; and 3 cubes will be left over.

EXTEND

1. ***What other questions about building and filling boxes occurred to you? What have you wondered about?*** Encourage student inquiry. Here are sample student queries.

   Why don't the large boxes look very realistic?

   Are there patterns that we can use to predict how many different rectangular prisms there are with a certain volume?

   Do boxes with the same volume have the same surface area? If they don't, what's the biggest surface area a box can have for the volume it has?

2. Have students write problems about building or filling boxes. One interesting problem type to introduce if students don't is this: ***A box is filled with exactly 30 cubes. The height of the box is 3 cm. What is the length and width of the box?*** Students should have no trouble determining that there are 10 cubes in a layer. They may be intrigued by the discovery that it is not possible to determine the box's length and width. Possible dimensions, in centimeters, are 1×10, 10×1, 2×5, and 5×2.

ANSWERS

2. Several answers are possible. Two solutions are 2 layers of 4 cubes, and 1 layer of 8 cubes.
3. Many answers are possible. Two solutions are 3 layers of 4 cubes, and 2 layers of 6 cubes.
4. Because 11 is a prime number, no set of numbers other than 11, 1, and 1 can be used as the dimensions of the box. The two possible solutions are 1 layer of 11 cubes, and 11 layers of 1 cube.
5. Many answers are possible. Two solutions are 6 layers of 4 cubes, and 4 layers of 6 cubes.
6. Many answers are possible. Two solutions are 6 layers of 6 cubes, and 4 layers of 9 cubes.
7. 7 layers
8. This is impossible. No even number is a factor of 27.
9. 18 layers

Stack It Up

Name:

Build boxes by making layers of cubes.

EXPLORE

1. Open **Stack It Up.gsp.** Go to page "Build a Box."
   You will drag points *L*, *W*, and *H* to make boxes.

2. Make a box with exactly 8 cubes. Record how many:
   cubes in each layer ___________
   layers ___________

3. Make a box with exactly 12 cubes. Record how many:
   cubes in each layer ___________
   layers ___________

4. Make a box with exactly 11 cubes. Record how many:
   cubes in each layer ___________
   layers ___________

Go to page "Two Boxes." Now you will build two different boxes, each with the same number of cubes.

5. Make two boxes, each with exactly 24 cubes. Record in the table.

	Cubes in a Layer	**Layers**	**Total Cubes**
Box 1			
Box 2			

6. Make two boxes, each with exactly 36 cubes. Record in the table.

	Cubes in a Layer	**Layers**	**Total Cubes**
Box 1			
Box 2			

EXPLORE MORE

Answer these questions. Tell about your thinking. If you want, go back to page "Build a Box" and use the model.

7. Christa used exactly 42 cubes to build a box. Each layer had 6 cubes. How many layers did the box have? __________

8. Marta is making a box with exactly 27 cubes. She wants to use an even number of cubes in each layer. How can she do that?

____

____

____

____

9. Roberto has 93 cubes. How many layers high can he build the box if 5 cubes fill each layer?

____

____

____

____

Perfect Packages: Surface Area and Volume

ACTIVITY NOTES

INTRODUCE

Project the sketch for viewing by the class. Expect to spend about 10 minutes.

1. Open **Perfect Packages.gsp.** Go to page "Box and Net."
2. Explain, ***Today you'll look at two different ways of describing the size of a package that is a rectangular prism. On the left side of the sketch you see a package that has a volume of 6 cubic units. On the right side this box has been unfolded into a net and the area of the net corresponds to the surface area of the package. What is the surface area of this package?*** Allow students to propose answers and encourage them to describe how they calculated their answers. It may be useful to point to the different-colored rectangles on the net and relate them to the sides of the package. ***As you can see, the surface area and the volume are not usually the same. Sometimes, depending on the purpose of the package, people want to make a box that uses as little surface area as possible while having a certain volume. Can you think of an example of when this might happen?***
3. Show students how to drag the sliders to change the length, width, and height. Ask, ***Can someone tell me how I could make a package that has the same volume as this one, but a different shape?*** Use the sketch to illustrate students' examples. These might include $1 \times 1 \times 6$. Ask students whether a $3 \times 1 \times 2$ package should be considered to have the same shape as a $2 \times 3 \times 1$ package. Opinions may differ, but let students know that this activity assumes they are the same shape. Ask, ***Do you think that all the packages with a volume of 6 cubic units have the same surface area? This is the first question you'll be exploring today.***

DEVELOP

Expect students at computers to spend about 25 minutes.

4. Assign students to computers and tell them where to locate **Perfect Packages.gsp.** Distribute the worksheet. Tell students to work through step 7 and do the Explore More if they have time.
5. Let pairs work at their own pace. As you circulate, here are some things to notice.
   - Help students find shorter ways than counting to calculate the surface area of the box.
   - Help students develop strategies to make sure they have found all possible ways of creating a box with a given volume. This might

involve making a table or listing all the factors of the value of the volume.

- Encourage students who finish quickly to investigate dimensions that are not whole number values (in the Explore More).

SUMMARIZE

Project the sketch. Expect to spend about 10 minutes.

6. Open **Perfect Packages.gsp** and use it to support the discussion. Invite students to share their strategies for finding the surface area of a package. If students bring it up, you might add to the projected sketch the surface area calculation described in the Extend section.

7. Ask students to talk about which packages had the least and greatest surface areas for a given volume. Students should notice that the least surface area occurs when the dimensions of the package are as equal as possible (that is, when the package is as cubic as possible), and that the greatest surface area occurs when two of the dimensions are very small and the third dimension is very large (that is, when the package is as long and thin as possible). Encourage students to consider dimensions that are not whole numbers. In this case the least surface area will occur when the dimensions are all the cube root of the volume.

EXTEND

1. Use Sketchpad's calculator to write a general equation for the surface area of the package (such as 2∗*length* + 2∗*width* + 2∗*height*, or 2∗(*length* + *height* + *width*)). Change the value of one of the dimensions and investigate how both the volume and the surface area change.

2. Discuss how the relationship between volume and surface area might change for prisms that don't have right angles, or for other solids. If students already know that a circle gives the maximum area for a given perimeter, you can discuss how a sphere gives maximum volume for a given surface area.

ANSWERS

1. Surface area = 22 square units
2. 1 × 1 × 6 box: volume = 6 cubic units; surface area = 26 square units

   1 × 1 × 5 box: volume = 5 cubic units; surface area = 22 square units
3. The values for length, width, and height are interchangeable.

   2 × 3 × 4 arrangement: surface area = 52 square units

   2 × 2 × 6 arrangement: surface area = 56 square units

   1 × 4 × 6 arrangement: surface area = 68 square units

   1 × 3 × 8 arrangement: surface area = 70 square units

   1 × 2 × 12 arrangement: surface area = 76 square units

   1 × 1 × 24 arrangement: surface area = 98 square units
4. The 2 × 3 × 4 arrangement has the least surface area. The 1 × 1 × 24 arrangement has the greatest surface area. (On page "Decimal Values," using approximately 2.9 units for each dimension will produce the least surface area.)
5. a. 2 × 2 × 2 (surface area = 24 square units)
   b. 2 × 2 × 3 (surface area = 32 square units)
   c. 3 × 3 × 3 (surface area = 54 square units)
   d. 3 × 4 × 4 (surface area = 80 square units)
6. a. 1 × 1 × 8 (surface area = 34 square units)
   b. 1 × 1 × 12 (surface area = 50 square units)
   c. 1 × 1 × 27 (surface area = 110 square units)
   d. 1 × 1 × 48 (surface area = 194 square units)
7. To find the least surface area, make the dimensions as equal as possible (so the package is as close to a cube as possible). To find the greatest surface area, make two of the dimensions equal to 1, and the third dimension equal to the volume (to make as long and thin a package as possible).
8. The package with the least surface area for 5b (volume = 12) is when each dimension is approximately 2.3 units. The package with the least surface area for 5d (volume = 48) is when each dimension is approximately 3.6 units.

The greatest surface area for each volume is when two dimensions are 0.1 unit and the third dimension is very large, but these no longer represent realistic packages.

9. Answers will vary. Sample answers: 1 × 1 × 5.5 (volume = 5.5 cubic units); 1 × 1.6 × 4 (volume = 6.4 cubic units); 2 × 2 × 2 (volume = 8 cubic units)

Perfect Packages

Name:

In this activity you'll investigate how changing the dimensions of a package (a rectangular prism) affects volume and surface area. Of course, you'll be most interested in the biggest and smallest packages you can make!

EXPLORE

1. Open **Perfect Packages.gsp** and go to page "Box and Net."

   If needed, change the dimensions to match those pictured at right, so the volume is 6 cubic units. What is the surface area?

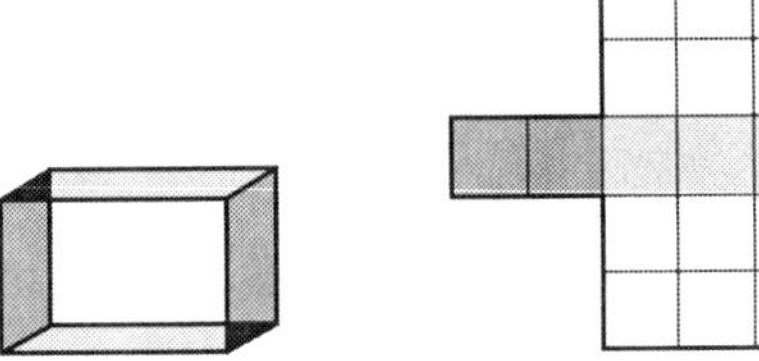

2. Can you find a package that has the same volume, but a different surface area? Can you find a package with the same surface area, but different volume? Drag the length, width, and height sliders to see whether you can come up with examples of such packages. Describe what you found.

3. Find all the ways 24 cubes can be arranged into a package. For each arrangement, write the dimensions and surface area in the table.

Length	Width	Height	Volume	Surface Area
			24 cubic units	
			24 cubic units	
			24 cubic units	
			24 cubic units	
			24 cubic units	
			24 cubic units	
			24 cubic units	
			24 cubic units	

Perfect Packages

continued

4. Which of your arrangements has the least surface area? Which has the greatest?

5. For each given volume, which package has the least surface area?
   a. Volume = 8 cubic units
   b. Volume = 12 cubic units
   c. Volume = 27 cubic units
   d. Volume = 48 cubic units

6. For each given volume, which package has the greatest surface area?
   a. Volume = 8 cubic units
   b. Volume = 12 cubic units
   c. Volume = 27 cubic units
   d. Volume = 48 cubic units

7. For any given volume, describe how you would find the packages with the least and greatest surface areas.

EXPLORE MORE

8. Go to page "Decimal Values," where you can adjust the dimensions to the nearest tenth of a unit. See whether you can find packages whose surface areas are less than those you found in step 5, or greater than those you found in step 6. Describe what you found.

9. Suppose you have a surface area of 24 square units. Can you make packages with different volumes? If so, provide two possibilities.

LaVergne, TN USA
10 April 2011
223362LV00001BB/1/P

9 781604 402278